CAMBRIDGE PHYSICAL SERIES.

GENERAL EDITORS:—F. H. NEVILLE, M.A., F.R.S.
AND W. C. D. WHETHAM, M.A., F.R.S.

MECHANICS

ARCHIMEDES

MECHANICS

BY

JOHN COX, M.A., F.R.S.C.

HONORARY LL.D., QUEEN'S UNIVERSITY, KINGSTON;
MACDONALD PROFESSOR OF PHYSICS IN MᶜGILL UNIVERSITY, MONTREAL;
FORMERLY FELLOW OF TRINITY COLLEGE, CAMBRIDGE.

CAMBRIDGE:
AT THE UNIVERSITY PRESS.
1904

CAMBRIDGE
UNIVERSITY PRESS

University Printing House, Cambridge CB2 8BS, United Kingdom

Published in the United States of America by Cambridge University Press, New York

Cambridge University Press is part of the University of Cambridge.

It furthers the University's mission by disseminating knowledge in the pursuit of education, learning and research at the highest international levels of excellence.

www.cambridge.org
Information on this title: www.cambridge.org/9781107676237

© Cambridge University Press 1904

First published 1904
First paperback edition 2014

A catalogue record for this publication is available from the British Library

ISBN 978-1-107-67623-7 Paperback

TO

ERNST MACH, Ph.D.,

PROFESSOR IN THE UNIVERSITY OF VIENNA,

WHOSE GENIUS HAS ILLUMINATED

THE HISTORICAL AND PHILOSOPHICAL DEVELOPEMENT OF

MECHANICS AND MANY OTHER BRANCHES OF

PHYSICAL SCIENCE,

THIS BOOK IS GRATEFULLY INSCRIBED.

PREFACE.

IT is a common complaint that though the principles of Mechanics are the simplest and the earliest to be discovered in the whole range of Science, and moreover are directly illustrated in almost every act of our lives, more difficulty is found in giving beginners a real grip of them than with any other branch of Physics.

This I attribute largely to the way in which the text-books deal with the subject. The student usually opens the book upon a chapter in which such leading concepts as matter, force, mass, particle, rigid body, smooth body are treated in definitions of a line or two each, before he sees any reason for their introduction at all. He is probably warned that philosophers are not agreed about the nature of matter; that motion is purely relative; that force is a misleading idea borrowed from our muscular sensations and better got rid of; and that no such things as mathematical particles, rigid bodies and smooth bodies exist in nature. He naturally concludes that Mechanics is an abstruse subject having nothing to do with realities or common sense.

The second chapter plunges him into the mathematical study of motion in the abstract. Here he struggles with variable velocity and acceleration, and the kinematic formulae; and is lucky if he is let off without a discussion of motion in a circle and in a cycloid, simple harmonic motion, and the parabola. To his previous confusion he adds the conviction that this is only another branch of the pure mathematics he has hitherto found so little use for.

At last there is a chapter on the Laws of Motion, so inadequately treated that he oftens ends by believing that they were made up by Sir Isaac Newton, the author, so far as he is aware, of the whole subject. The rest of the book is too often merely geometrical and trigonometrical gymnastics.

In recent years many text-book writers have attempted to break away from this mischievous tradition. Some have tried to rewrite the whole subject from the latest point of view of Energetics. But this is surely to begin at the wrong end. According to the biologists the bodily developement of the individual is an epitome of the developement of the race. Is not this a hint that the historical method is the natural way of attacking a subject of study? Others have sought to discard the idea of force, and speak only of mass-accelerations. "Naturam expellas furca." It is rarely indeed that they manage twenty pages without getting back to the old point of view. With proper caution the use of this concept is as valuable as it is historically right and inevitable. Still others have set the student to rediscover the subject for himself by experiment. But this wastes too much time on mere manipulation, and leaves the student's knowledge in mid air, unrelated to all that has gone before him in the course of actual discovery. It seems a pity that he should close the book without a glimmering of personal interest in his predecessors, the great investigators, and forego the insight into philosophic and scientific method which a study of the developement of Mechanics evokes insensibly and unawares.

No claim for originality can be made for this book. And yet I find it difficult to make detailed acknowledgement of obligations, for it embodies the system of teaching Mechanics at which I have arrived after thirty years' experience, and it is no longer possible to say where certain illustrations and ways of putting things have come from, or whether (less probably) they were devised by myself. A desire to shew whence the general plan has arisen must be my excuse for a personal digression.

After learning and teaching Mechanics for ten years on the traditional system described above, I was called on, as a lecturer under the Cambridge University Extension Scheme, to explain the principles to audiences without any previous mathematical training, but often composed of engineers, plumbers, and other workmen who had derived excellent practical notions on the subject from their experience. Obliged thus to recast the subject in my own mind, I found it possible to present all the main principles with the aid of ordinary arithmetic and the simplest geometrical diagrams. At this stage Sir Robert Ball's admirable lectures on *Experimental Mechanics* gave me great assistance. My experience with these popular audiences reacted with advantage on my teaching with classes in the University, and fired me with the ambition to write a text-book on Mechanics. But a sight of Sir Oliver Lodge's excellent *Mechanics* in Chambers' series put an end to this wish for a time. Some ten years ago I stumbled on the first German edition of Professor Mach's *Die Mechanik in ihre Entwickelung*. I am ashamed to say that this fascinating book was my first introduction to the historical developement of a subject I had taught so long. Since then my teaching has been based more and more on the lines laid down by Mach, and as I have found it impossible to induce ordinary students to read the original, even when translated, I recurred to the idea of writing a text-book which should yet be based on Mach's method. The present book is the result. In producing it I have kept in view the following aims :—

(1) First and throughout, to make a text-book of mechanical principles, avoiding as far as possible merely mathematical difficulties, and reserving those that could not be avoided for separate treatment in the later parts of the book ; but never shirking them where necessary.

(2)   To develope the principles in their historical order, starting from real problems, as the subject started, shewing how the great investigators attacked those problems, and only introducing the leading concepts as they arise necessarily and naturally in the course of solving them.

(3)   To bring out incidentally the points of philosophic interest and the method of science.

(4)   To appeal constantly to experiment, as far as possible in the original form, for purposes of verification in the early part of the subject, leading up to an experimental course limited to the most important practical applications.   In this way a good deal is included in the later chapters which does not usually find a place in elementary text-books.

(5)   To interest the student in the personality of the great pioneers, and if possible induce the habit of referring to original sources.

(6)   Not to overload the text with masses of examples; to give only so many that every student should work them all; and to select these so as to bring in useful and interesting physical constants, and make them as direct as possible, discarding all those in which the amount of mechanical principle involved is a mere drop to an intolerable deal of pure mathematical exercise.   Any intelligent teacher can multiply examples of a given type, where this is necessary, either from his own invention or from the numerous extant collections.   Unintelligent teachers have no business with Mechanics.

It will be obvious that the Book could never have been written but for Mach's *Mechanik*.   It is indeed only a poor and incomplete abridgement of Mach's work intended for students.   I trust that every teacher into whose hands it may fall, and many students, will be driven by it to the original.   It is impossible for me to express my personal obligation to Professor Mach in connection with this subject, or the respect and gratitude I feel for such a master for enlightenment and inspiration in this as in many other branches of Physics.

Further acknowledgement must be made of help from Clerk Maxwell's *Matter and Motion*, Sir Oliver Lodge's *Pioneers of Science*, Mr W. W. R. Ball's *Essay on Newton's Principia*, Professor Wright's *Mechanics*, and Glazebrook and Shaw's *Practical Physics*, to which many of the experiments described in the latter part of the book are due. The *Dynamics* of Principal Garnett must be specially mentioned, for it was in reading the proof-sheets of that work that I first learned to connect the familiar formulae with practical facts.

My especial thanks are due to Mr F. H. Neville, one of the Editors of the Cambridge Physical Series, not only for his extreme care in revising the book for the press, but for many most valuable criticisms and suggestions that have led to important improvements.

The course, as laid down, has been tested with classes at McGill University, and has, I think, proved interesting and intelligible. Many of the illustrations are from photographs of actual apparatus used here. Whether the book will be found adapted to students anywhere else is a subject of much misgiving. It aims at no particular examinations, and may be too revolutionary to find favour in schools. Its execution is probably faulty enough. But that something of this nature would be a more worthy treatment of the subject for university students, even for the ordinary degree, than the present jejune versions of Varignon's *Statics* and mathematical exercises on kinematics, I have no doubt at all. Until Mechanics is clad in its historical flesh and blood, it will remain the dull and tiresome subject that has convinced so many generations of students that an abysmal gulf separates theory from practice.

J. C.

MONTREAL,
April, 1904.

CONTENTS.

BOOK I.

THE WINNING OF THE PRINCIPLES.

BOOK II.

MATHEMATICAL STATEMENT OF THE PRINCIPLES.

BOOK III.

APPLICATION TO VARIOUS PROBLEMS.

BOOK IV.

THE ELEMENTS OF RIGID DYNAMICS.

PLATES.

BOOK I.

THE WINNING OF THE PRINCIPLES.

INTRODUCTION.

By Mechanics is understood nowadays the science, or organized body of knowledge we possess concerning the conditions of rest or motion of the objects about us.

How did we come by it? The word itself means "contrivances," and gives a hint that the science arose from the devices which were found helpful in lifting weights and moving objects to satisfy practical needs. Long before there was any collection of rules, much less a science of Mechanics, the advantages of the traditional "Mechanical Powers"—the Lever; the Wheel and Axle (or continuous lever); the Pulley (or travelling lever); the Inclined Plane; the Wedge (or double inclined plane); and the Screw (or continuous inclined plane)—were known. The great monuments of antiquity, like the Pyramids, could hardly have been raised by the labour of unaided hands. As a matter of fact, rude implements of the kind have been found in ancient graves, and the Egyptian and Assyrian records contain pictorial representations of such appliances.

The transition to what may be properly called science takes place when, for example, instead of the practical knowledge that a great weight may be lifted by a small one with the aid of a crowbar, a principle, or rule, is discovered, which tells us what must be the lengths of the arms of the crowbar in order that a certain small weight may lift a given large weight. This is a step of immense importance. For once it is made, a craftsman can save innumerable mistakes, with their consequent loss of time and risk of injury, by calculating beforehand what weight or length of lever to employ; and he can communicate to others what he has learned from his own experience. Thus both Design and the

Dissemination of Knowledge become possible, as when, during the siege of Kimberley, a mining engineer constructed out of a steel axle, ten feet long, a gun capable of replying to the Boer artillery, gathering his information out of some back numbers of the engineering magazines, which happened to be in his possession.

Statics, the part of the subject which deals with Equilibrium, being simpler and far more directly concerned with the Mechanical Powers, took its rise much earlier than Dynamics, the science of motion as produced by force. It begins with the clearing up of the Principle of the Lever by Archimedes, to whom is also due the fundamental principle of Hydrostatics. Strange as it may seem, these are the only important contributions to Physical Science of the ancient world up to the middle of the sixteenth century, when, after an interval of eighteen hundred years, Stevinus of Bruges attacked the study of Statics again, this time by way of the Inclined Plane.

From this time onwards a series of great investigators,— Galileo (1564—1642), Huyghens (1629—1695), and Newton (1642—1727)—laid the foundations of the Dynamics of a single particle, and bodies that could be treated as such. D'Alembert (1743) gave a general principle by which Newton's ideas could be applied to the complicated case of solid bodies consisting of innumerable particles.

All that remained to be done was to employ the highest developments of Pure Mathematics in working out the consequences of the principles already discovered; save that towards the middle of the nineteenth century the importance of the doctrine of Energy came to be more and more recognized, and the great generalization known as the Conservation of Energy, dimly foreshadowed in the Principle of Virtual Work and the Scholia to Newton's Third Law of Motion, was finally established and extended from Mechanics to all departments of Physics.

We shall begin with Archimedes and the Principle of the Lever. It is instructive to examine this case in some detail, not only for its historical interest, but because it is an admirable example of the way in which Physical Science has developed.

CHAPTER I.

Δὸς ποῦ στῶ, καὶ τὴν γῆν κινήσω.

1. ARCHIMEDES (287—212 B.C.), the greatest mathematical
and inventive genius of antiquity, was born at Syracuse, and completed his education at Alexandria under Conon, in the Royal
Schools of the Ptolemies, of which Euclid had been an ornament
fifty years earlier. The stories of the crown of Hiero, the burning
mirrors, and his slaughter at the end of the siege in spite of
Marcellus' orders are well known.

2. Every one knows that a small force applied to one end of
a lever, a long way from the point of support, or fulcrum, will
overpower a much larger force applied near the fulcrum. The
problem is to find a rule connecting the forces and their distances
from the fulcrum when they just balance.

Those who consult the treatise of Archimedes περὶ ἐπιπέδων
ἰσορροπικῶν ἢ κέντρα βαρῶν ἐπιπέδων, will be struck by two
things. He deals entirely with weights, not introducing the
general notion of force; and he is so deeply imbued with the
methods of the ancient geometers that he tries to cast his proofs
of physical propositions into the same form. His very real grip
of the Principle of the Lever must have been, in his case as well
as that of the craftsmen of his age, a direct result of experience.
Yet he finds a satisfaction in reducing it to the already familiar
array of Axioms, Proofs by Reductio ad Absurdum, and the
geometrical theory of Proportion.

He begins by laying down the following Axioms:

(1)   Equal weights placed at equal distances from the point of support balance.

(2)   Equal weights placed at unequal distances do not balance, but that which hangs at the greater distance descends.

Then follows a proof by *reductio ad absurdum* that in the case of unequal weights balancing at unequal distances, the greater weight must be at the shorter distance.   Before advancing to the actual numerical law connecting the weights and the distances, he now lays down three propositions to shew that the Centre of Gravity of any number of equal weights, odd or even, equally spaced out along a bar, must be at the middle point of the bar. Observe that in these propositions it is clear that he conceives of the centre of gravity as a point such that, if it be supported, the weights will balance about it.

He is now able to give a beautifully ingenious proof of the general principle of the Lever, viz. that for equilibrium the weights must be inversely proportional to the distances; first for the case of commensurable, and then for incommensurable weights.   As a historical curiosity, we shall give the former in his own words.

3.   *Proposition*.   Magnitudes whose weights are commensurable will balance if they are hung at distances which are inversely proportional to their weights.

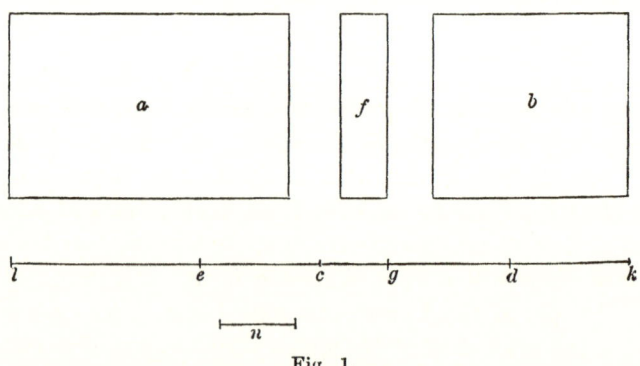

Fig. 1.

Let *a*, *b*, be commensurable weights.   Let *ed* be any distance, and let *dc* be to *ce* as *a* is to *b*.   It has to be proved that the centre of gravity of the magnitude composed of both a and b, placed at e and d respectively, is the point c.

Since dc is to ce as a is to b, and a is commensurable with b,

∴ dc is commensurable with ce, a straight line with a straight line.

∴ there must be some common measure of dc, ce.

Let it be n; and take dg, dk, on each side of d, equal to ce, and el equal to dc.

Since $$dg = ec,$$

∴ also $$dc = eg,$$

∴ also $$le = eg,$$

∴ lg is double of dc, and gk of ec.

∴ n will measure both lg and gk, since it measures their halves.

And since dc is to ce, as a is to b, and lg is to gk, as dc is to ce, for they are the double of each, ∴ lg is to gk as a is to b.

Now let a be the same multiple of a magnitude f, that lg is of n.

Then a is to f as lg is to n.

But kg is to lg as b is to a.

∴ ex aequali, b is to f as kg is to n.

∴ kg is the same multiple of n that b is of f.

But it has been shewn that a is also a multiple of f.

∴ f is a common measure of a and b.

If therefore lg be divided into parts equal to n, and a into parts equal to f, the parts of lg, each equal to n, will be the same in number as the parts of a, each equal to f.

∴ if to each of the parts of lg there be applied a magnitude equal to f, having its centre of gravity in the middle of the part, all the magnitudes will together be equal to a, and the centre of gravity of the magnitude composed of all of them will be e, for they are equal in number on opposite sides, since $le = eg$.

Similarly it can be shewn that if to each of the parts of kg there be applied a magnitude equal to f, having its centre of gravity in the middle of the part, all the magnitudes will together be equal to b, and the centre of gravity of the magnitude composed of them all will be the point d.

a has therefore been placed at e, and b at d, and there are now an even number of equal magnitudes, placed in a straight line, whose centres of gravity are equally distant from each other.

It follows that the centre of gravity of all the magnitudes together is the point of bisection of the straight line in which the centres of gravity of the magnitudes lie.

∴ since $le = cd$, and $ec = dk$, the whole lc = the whole ck.

∴ the centre of gravity of the whole is the point c.

∴ if a be placed at e, and b be placed at d, they will balance about the point c. Q. E. D.

4. But little advance was made on this cumbrous proof for 1800 years, when Stevinus of Bruges (1548—1620 A.D.) gave it the following interesting form.

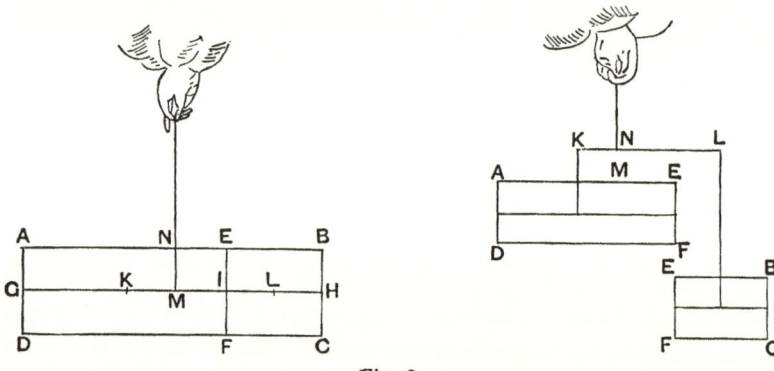

Fig. 2.

Consider a uniform column, AC, suspended by its middle point M, so that it will balance. Imagine it divided at EF into two parts whose middle points will be K and L. Then the weights of AF, EC are proportional to GI and IH, and may be supposed collected at K and L.

It is easily seen that $KM = IL$ and $ML = IK$.

∴ the greater weight AF is to the smaller weight EC as the longer arm ML is to the shorter arm MK.

The second figure shews that the weights may be hung at any depth below the bar, and that any equal weights may be substituted for the parts of the bar.

This form of the proof was adopted, with a slight modification, by Galileo (1638) and in modern times by Lagrange.

5. In this proof Archimedes and his successors apparently
Criticism of   evolve a physical truth by geometrical methods
the proof.   from certain axioms which are assumed as self-
evident apart from experience. But can this be possible?

Take the first axiom. It seems perhaps self-evident that equal
weights at equal distances from the fulcrum must balance, from
the mere symmetry of the figure. The ancient philosophers,
steeped in the methods of logic and geometry, base such cases of
symmetry on what was called the "Principle of Sufficient Reason."
No motion can take place, because there is no reason why the
balance should descend on one side more than the other. "But
we forget in this that a great multitude of negative and positive
experiences is implicitly contained in our assumption; the negative,
for instance, that unlike colours of the lever arms, the position of
the spectator, an occurrence in the vicinity, and the like, exercise
no influence; the positive, on the other hand (as it appears in the
second axiom), that not only the weights, but also their distances
from the supporting point are decisive factors in the disturbance
of equilibrium" (Mach, *The Science of Mechanics*).

The secret of the immense and rapid development of natural
knowledge in modern times lies in the deliberate and faithful
ransacking of nature for her facts, since the time of Francis
Bacon's *Novum Organon*. Natural processes can only be learned
from experience; they cannot be extracted from the meanings of
words or the canons of logic, after the manner of the ancient
world, except in so far as these themselves enshrine the results of
direct experience, hereditary or personal. Why should not the
position of the sun affect the balancing of an equal-armed, equally-
weighted lever, so that it should be horizontal at noon and mid-
night, and its eastern limb dip in the morning, its western in the
evening? Nothing but experiment can teach us that the sun has
no effect in this case.

Let a wire pointing to magnetic north be stretched horizontally
over a compass needle, and let an electric current be sent through
the wire from South to North. Everything is symmetrical. It
might seem an axiom in accordance with the principle of sufficient
reason that the magnet will remain at rest. Yet Oerstedt discovered in 1821 that the north pole of the magnet will certainly
turn to the west.

In the Proposition Archimedes is seeking to reduce the general and more unfamiliar case of a lever with unequal arms to the case of the equal-armed lever, which was already so familiar to him that the knowledge of it seemed instinctive or axiomatic. Think of any case of scientific explanation, and you will see that this is all that is accomplished,—the reduction of unfamiliar cases to those already familiar. Newton discovers that the moon in her orbit drops towards the earth according to the same law that is familiar in the falling apple; the motion of the moon is *explained*, but not that of the apple.

But did Archimedes succeed? It seems unlikely that a method which, apart from experience, fails to justify his first axiom, can possibly lead him to the numerical law of the lever. Closely scrutinized, the fallacy appears. He assumes that a number of weights spaced out along one arm of a lever will have the same turning effect about the fulcrum, as if they were all collected at their centre of gravity; whereas what he has proved from his axioms in the preliminary propositions is that they will balance about their centre of gravity, if it be supported.

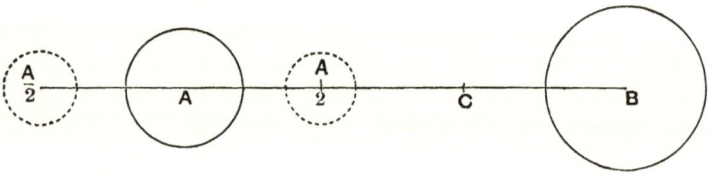

Fig. 3.

Suppose it had been a question, not of balancing, but of the resistance experienced upon attempting to set the lever AB rotating rapidly about C. We require to find the law connecting unequal weights at unequal distances, so that they may offer the same resistance. May we substitute for A, two weights each equal to $A/2$, placed symmetrically about A? Certainly not. But it is only experience that tells us we may do in a question of balancing, or centres of gravity, what we may not do when it concerns moments of inertia and rotation. But Archimedes is not aware that he has made the step, because he has been busied with the principle of the centre of gravity, which is in fact equivalent to the principle of the lever. He would not have attempted the proof, if he had not first discerned the principle directly, and the fact of experience, once discerned, has as great an authority in the general case, as in the simple.

But the achievement of Archimedes is not in vain, for it brings into vivid relief the connection between the general case of unequal arms, and the special and more familiar instance, when the arms are equal; and we derive a satisfaction from our insight into their consistency.

CHAPTER II.

6. WHAT the early investigators learned from their own experience and that of the craftsmen, the student of to-day can only grasp with equal vividness by experimenting for himself.

The Principle to be verified is this:—

Let AB be a light rod supported at C, and let weights, P and Q, be hung at A and B. Then for equilibrium

$$P : Q :: BC : AC,$$

or more conveniently,

$$P \times AC = Q \times BC,$$

i.e., the products of the weights by the arms at which they hang are the same.

7. *Experiment* 1. Take a graduated rod, say a 30-centimeter length of a meter rod, from which two scale-pans can be supported by loops of fine wire. Suppose the weight of each scale-pan is made up to 50 grams by adding lead shot. Place a 50-gram weight in each, and balance the rod over any sharp edge, such as a small metal or glass prism, on a corner of a table, so as to allow the scale-pans to hang below, one at each end, and each of them 15 cm. from the prism at the centre. Keeping one of the scales unchanged, find where the other scale must hang in order to balance about the centre, when the weight in it is increased by 20 gms. at a time up to a total weight of 250 gms.

Make a table as follows :—

Products

100 gms. (including scale) balance 100 gms. at a_1

„	„	„	„	120	„	a_2
„	„	„	„	140	„	a_3
				⋮		
„	„	„	„	250	„	

a_1, a_2, &c. being the observed distances from the centre of the rod at which the second scale has to be hung to secure a balance.

The rod may not be quite uniform, so that a_1 may not be exactly 15 cms. But work out the products of the weights and distances on the right-hand side, and see whether they are as nearly equal as the accuracy of your method would reasonably lead you to expect.

8. Stevinus, as Archimedes 1800 years before him, is thinking always of real weights. Even when the pull is to be exerted upwards, as in supporting the rod at C, it is applied, as his figures

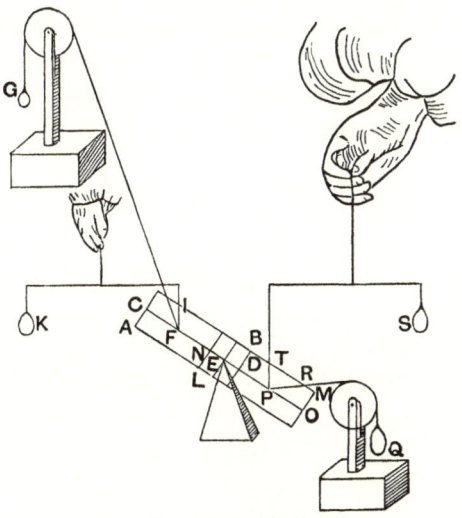

Fig. 4.   (From Stevinus.)

shew, by means of a string carried over a fixed wheel, or pulley, with the proper weight hanging from the other end.

If R is this sustaining weight, we may verify that
$$R = P + Q.$$

Experiment 2. But it is more convenient to employ a Spring Balance, which determines weights by the amount to which they can pull out a spiral spring fastened at one end.

Attach such a spring balance to the centre of the rod, having first verified the readings of the balance by testing it with a set of standard weights, and made a table of errors.

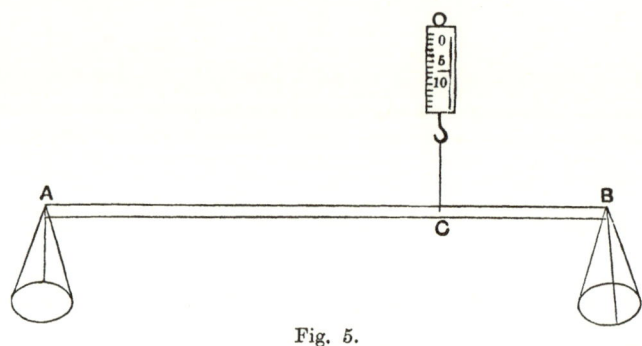

Fig. 5.

Repeat several of the above experiments, noting the total weight supported. Weigh the rod without the scale-pans, and subtract its weight from each of the total weights. See if the remainders are not in every case equal to the sum of the weights suspended from the rod.

9. *Experiment* 3. Attach a small object, such as a metal clamp, to one end of the rod, and balance it without scale-pans. The point over the edge of the prism must be the centre of gravity of the rod and clamp together, and their joint weight may be supposed to act at it.

Hang a scale-pan with a known weight near the other end, and find where the whole balances. Hence calculate the weight of the rod and clamp together, and verify your result on the spring balance.

10. In these experiments any of the weights concerned may

Force.   be replaced by a pull or push applied either by a spring balance, or direct muscular effort, which would just sustain the weight. In such efforts we speak of exerting a force, and it is convenient to introduce this term at once, though the general idea of Force, as *anything which changes or tends to change a body's state of rest or motion*, belongs to Newton's time, half a century after Stevinus.

11. Our experiments shew that if two forces be applied,

Moment of a Force about a point.

perpendicularly, to the ends of a rod capable of turning about a fixed point, or fulcrum, they will balance provided the product of each force into the lever arm at which it acts is the same for both. In the case we have considered the lever arms are the shortest, or perpendicular, distances from the fulcrum to the lines of action of the forces.

Leonardo da Vinci (1452—1519), the famous painter, engineer, and investigator, recognized that this is the essential condition in all cases, even when the forces act obliquely. He says, for example : We have a bar AD free to rotate about A, and suspended

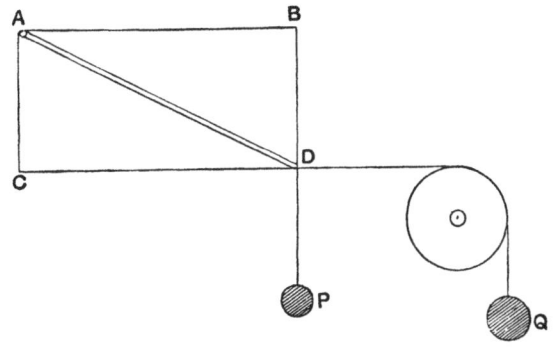

Fig. 6.

from the bar a weight P, and suspended from a string which passes over a pulley, a second weight Q. What must be the ratio of the forces that equilibrium may obtain ? The lever arm for the weight P is not AD, but the "potential" lever AB. The lever arm for the weight Q is not AD, but the "potential" lever AC.

Professor Mach suggests that Leonardo may have reached this idea in some such way as this. Consider a string laid round a

pulley, and subject to equal tensions on both sides. *EF* will
be a plane of symmetry, and
we see that equilibrium will
subsist.   But we also note
that the only essential parts of
the pulley are the two rigid
radii, *AB, AC,* which determine
the form of the motion of the
points of application of the two
strings. If nails were driven
through the string at *B* and
C, the rest might be cut away
without disturbing equilibrium.
Hence, in Fig. 6, the lever arm
for the right-hand force is not
AD, but the "potential" lever
AC.

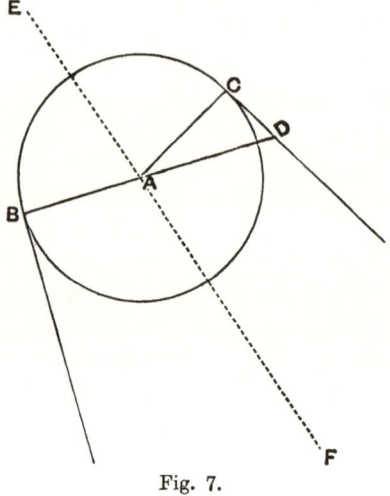

Fig. 7.

However this may be, it was recognized that the *torque,* or
tendency of a force to turn a body about a pivot, depended only
on two things, the magnitude of the force, and the perpendicular
distance from the pivot to its line of action, and that two forces
had equal torques, if for each the product of the force by the
perpendicular distance of its line of action from the fulcrum was
the same. This product is therefore the measure of the torque,
or tendency of a force to turn a body about a pivot.

It is evidently high time to introduce a single word for the
very important but cumbrous expression "product of a force into
the perpendicular distance from the point to the line of action
of the force." It is called the *Moment* of the force about the
point.

Definition. The *Moment* of a Force about a point is the
product of the Force by the perpendicular distance of the point
from the line of action of the force.

We can now state the Principle of the Lever, including the
case of oblique forces, as follows:

Two forces acting on a lever will balance when their Moments
about the fulcrum are equal and opposite.

12. The Lever and the other Mechanical Powers were
Mechanical   employed to enable a small force to balance or
Advantage.   overcome a large weight or force. In this they
are said to afford " Mechanical Advantage." The mechanical
advantage is measured by the ratio of the large weight to the
force required to balance it. Tradition has fixed the use of the
terms Power and Weight to indicate the force employed and the
resistance, whether weight, or pull, or push, overcome. This is
rather unfortunate, as Power has a definite and quite different
meaning in Dynamics.

The mechanical advantage of the Lever, then,

$$= \frac{\text{Weight}}{\text{Power}} = \frac{\text{Length of Power arm}}{\text{Length of Weight arm}}.$$

In the machines the weight moved is not always greater than
the power. When it is less, the power is said to act at mechanical
disadvantage.

13. Consider a lever with arms a, b, weights P, Q, and sup-
The different   ported at the fulcrum by a force $P + Q$ applied
kinds of levers. by a spring balance. It does not matter how the
three forces are applied at A, B, C, provided they have the proper

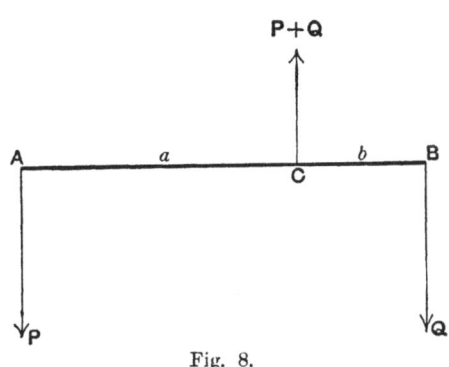

Fig. 8.

magnitudes. We may therefore regard either A or B as the
fulcrum, just as well as C.

(1) If the fulcrum is at C, between the power and the weight,
the lever is said to be of the *first class*, and there will be mechanical
advantage or disadvantage according as a is greater or less than b,
since $P \times a = Q \times b$.

Fig. 9.   Buckton 150-ton Testing Engine in the Laboratories of McGill University, Montreal.   The V-supports, of which there are two sets for use with different scales, are seen at A.   The beam is balanced when the 2000 lb. weight B is at the extreme right.   The specimen to be tested is held in jaws at C.   The pull is applied to the lower end by hydraulic machinery in the room below, and the weight B is shifted by the gear-wheel D so as to keep the beam balanced.   The tension at which the specimen yields, as well as continuously throughout the operation, is read on the scale.

Examples of levers of the first class are :—*Single levers.* A poker (lifting the coals by resting on the bars of the grate as a fulcrum); a crowbar; the shadoof, or pole and bucket, a device used in Egypt for raising water from the Nile; a Testing Engine.

Double levers. A pair of scissors; a pair of pincers.

(2) Regard B as the fulcrum, and $P + Q$ as the "weight"; the power P is on the same side of the fulcrum, but farther off. The lever is of the *second class*, and there is always mechanical advantage. The principle of the lever still holds good, for

$$(P + Q) \times BC = P \times BC + Q \times BC$$
$$= P \times BC + P \times AC$$
$$= P \times (BC + AC)$$
$$= P \times AB,$$

i.e. product of power and its arm = product of weight and its arm.

Examples of levers of this class are :—*Single levers.* The oar of a boat. (The broad blade, approximately fixed in the water, acts as fulcrum.) A door, when used to crack a nut in the hinge.

Double levers. A pair of bellows. A pair of nutcrackers.

(3) Regard B as the fulcrum, but P as the weight and $P + Q$ as the power. The lever is said to be of the *third class*, and there is always mechanical disadvantage. As above,

product of power and its arm = product of weight and its arm.

Examples :—*Single levers.* Most of the limbs of the body are of this class. Thus the forearm moves about the elbow-joint as a fulcrum. The power is applied (very obliquely too) by the biceps muscle. The mechanical disadvantage is very great, and the muscles must possess great strength; but this could not be avoided unless the human body were constructed so as to resemble an animated derrick, which would be awkward for locomotion and activity.

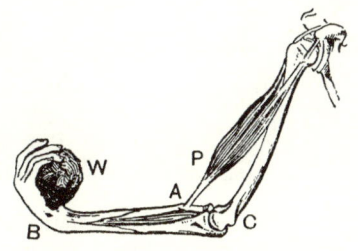

Fig. 10.

Double levers. A pair of sugar tongs. A pair of tweezers.

Fig. 10 *a*. A striking illustration of Levers of the Third Class is found in the
Hydraulic Scale of the Emery Testing Engine in the Testing Laboratory of McGill
University, Montreal.   The pressure is conveyed from the ram of the engine to the
drum above P_1 through copper pipes.   This is applied at the knife-edge P_1 to the
lever whose fulcrum is f_1.   The force is reduced by a series of levers of the third
class, and conveyed to the central weighing lever, and the deflection of the latter
is magnified by the upper lever.   The weights are applied automatically by raising
the four handles to the left.

14. A straight lever, working on a fixed fulcrum, can only
The Wheel and Axle. raise the weight to a height above the fixed fulcrum equal to the length of the short arm.
This difficulty is got over in the second of the mechanical powers, the Wheel and Axle. It consists of a wheel of large radius rigidly bolted to an axle of smaller radius. The weight is hung from a cord coiled on the axle. The power is applied to the large wheel by pulling on a cord coiled round its circumference, or by a handle projecting from its rim, as in the familiar device for raising water from wells.

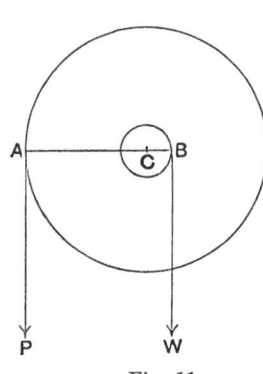

Fig. 11.   Fig. 12.

It is obvious that the condition of equilibrium is the same as for the lever

$$P \times AC = W \times BC.$$

In fact at any given instant the radii AC, CB form a straight lever. But as each radius moves out of position, the next takes its place. The wheel and axle may therefore be regarded as a *continuous* lever.

Another form of the wheel and axle is the Capstan, where the power is applied by handspikes, and the resemblance to a lever is still more obvious.

15. The simple Pulley is a wheel with grooved edge round
The Pulley. which a cord is passed and supported at one end. The power is applied to the other end. The weight is hung from the axle of the pulley.

2—2

At any instant the diameter ACB may be regarded as a lever with fulcrum at A. Hence

$$P \times AB = W \times AC,$$

or $\qquad \dfrac{P}{W} = \dfrac{AC}{AB} = \dfrac{1}{2}.$

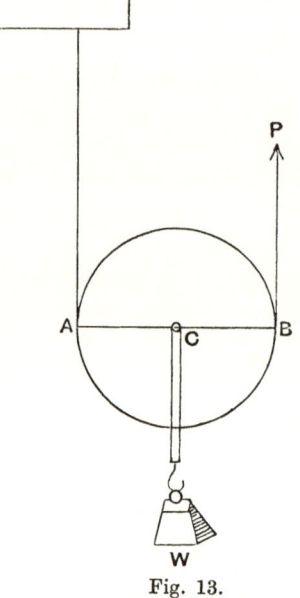

(Otherwise thus :—The tension of the string must be the same on each side, or else the pulley would turn. Hence the weight is supported by two pulls applied at A and B, each equal to P. Therefore $P = W/2$. This is clear enough if each string is held up by a man. It makes no difference, however, if one end, instead of being held by a man, is fastened to a fixed support.)

If the weight of the pulley is too great to be neglected, let it be w.

Then $\qquad P = \dfrac{W + w}{2}.$

Fig. 13.

As the pulley rises, fresh diameters take the place of AB, and since the fulcrum moves, we may regard the pulley as a *Travelling* lever.

16. A single moveable pulley only enables us to double the force at our disposal. By combining several pulleys we may increase the mechanical advantage to any extent. The following combinations are in common use, or interesting historically.

Systems of Pulleys.

(1) Archimedes' System. (Fig. 14.)

By the principle of the simple pulley the tension in each string is double that of the string next above it. The weight is double the tension of the last string. Hence if there be n moveable pulleys,

$$W \times 2^n = P,$$

and $\qquad \dfrac{P}{W} = \dfrac{1}{2^n}.$

(2) The Pulley Block. (Fig. 15.)

There are two blocks, each containing several pulley-wheels, or sheaves, on the same axle. The string is fastened to one of the blocks, and then carried round all the sheaves as in the figure.

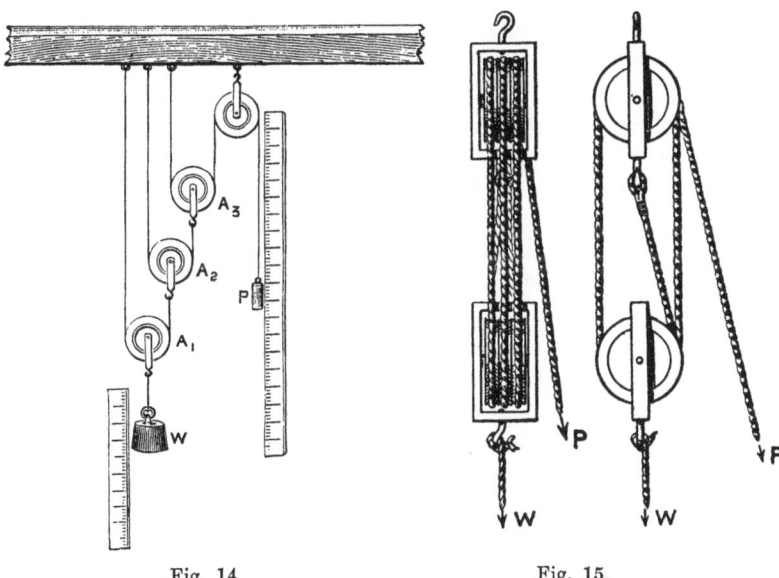

Fig. 14.   Fig. 15.

The tension of the string is the same throughout, so that the weight is supported by as many tensions each equal to the power as there are strings at the lower block. Count these, and let their number be n. Then

$$\frac{P}{W} = \frac{1}{n}.$$

If it be desired, allowance can be made for the weights of the pulleys as before.

EXAMPLES.

1. A pump handle is 3 ft. 8 in. long, and works on a pivot 4 in. from the end attached to the pump rod. What force is applied to the pump rod when the handle is pushed down with a force of 10 lbs. weight?

2. A safety valve consists of a circular hole, $\frac{1}{4}$ inch in diameter, closed by a plunger attached to a light horizontal hinged bar one inch from the hinge. A weight of 1 lb. slides on the bar. How far from the hinge must it be set if the steam is to blow off at 160 lbs. on the square inch ?

3. An oarsman weighing 180 lbs. pulls horizontally at the handle of an oar so as just to lift his weight from the seat. The stretcher against which his feet press is 16 inches below the level of his hands, and distant 2 ft. from the vertical through his centre of gravity. What is the force applied to the oar ?

4. If, in example 3, the rowlock is at one-quarter of the distance from the hands to the blade of the oar in the water, what propelling force could eight such oarsmen apply to the boat ?

5. Six men work a capstan using handspikes projecting 5 ft. 3 in. from the centre. The barrel on which the rope is coiled is 2 ft. 3 in. in diameter. What force must each man exert in order to raise a weight of a ton and a half ?

6. The rope of the simple pulley, Fig. 13, is carried over a fixed pulley and held by a man who supports himself by standing in the hook attached to the moveable pulley. What is the pull on the rope if the man weighs 180 lbs. ?

7. If there are four pulleys in the system of Archimedes, what force is required to support a weight of 2 cwt. (1) when the weight of the pulleys is neglected, (2) when each pulley weighs 8 lbs. ?

CHAPTER III.

THE CENTRE OF GRAVITY.

17. THE principle of the lever shews us that two weights rigidly attached to a light rod will balance if their moments about the fulcrum are equal and opposite, and that the fulcrum must be supported by a force equal to the sum of the weights. This principle may be generalised in two ways.

(1) Let another pair of weights be attached to the same rod. Then if their moments about the fulcrum are equal and opposite, they also will balance. It is a fact of experience that the presence of the one pair in no way interferes with the equilibrium of the other. The same is true for any number of pairs.

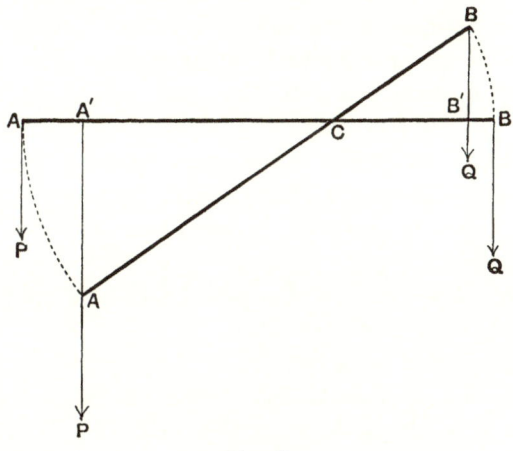

Fig. 16.

Hence any number of weights at different distances on a rod will balance provided that the sum of the moments on one side of the fulcrum is equal to the sum of the moments on the other side.

(2) The rod may be turned through any angle about the fulcrum, and yet equilibrium will subsist.

For by similar triangles

$$\frac{CA'}{CB'} = \frac{CA}{CB},$$

$$\frac{Q}{P} = \frac{CA}{CB} = \frac{CA'}{CB'},$$

$$P \times CB' = Q \times CA',$$

and the moments are still equal.

A rod so weighted that the sum of the moments on each side of the fulcrum is the same may be said to be mechanically symmetrical about the fulcrum.

It is clear that there may be as many such rods as we choose, all rigidly joined at the fulcrum, and yet the whole system will balance in any position about it.

Now the objects with which we deal in Mechanics consist of innumerable small parts, or particles, rigidly joined together and each possessing its own weight. Mechanical problems will be enormously simplified if we can find for any object the point about which it is mechanically symmetrical. For if this be supported by a force equal to the total weight of the object, equilibrium will subsist, since the object will certainly balance about this point. For many purposes we need no longer consider the myriads of small weights, but replace them by a single weight at this point. Hence the point is called the Centre of Gravity of the object.

Definition. The Centre of Gravity of a body is the point about which it will balance in all positions.

Two things should be noted:

(1) It is not sufficient that there should be mechanical symmetry in one direction, say right and left. A vertical rod will remain at rest however the weights are distributed on it, even though all of them should be above the fulcrum. For since all the perpendiculars from the fulcrum on the vertical lines of action of the weights are zero, the moments on each

side of the fulcrum are zero, and there is mechanical symmetry horizontally, but not vertically.

But if the rod be turned ever so slightly from the vertical, equilibrium is at once destroyed. If an object be found to balance about a point in more than one position, then it will balance in all positions, and the point of support must be the centre of gravity.

(2) There cannot be two centres of gravity for the same body, for if the body were turned so that the line joining the two centres was horizontal, the moments to the left and right could not be equal for both points at the same time.

18. *Experiment*. Find experimentally the centre of gravity of a flat board.

Bore two small holes near the rim, and suspend the board from a knitting-needle passed through one of them at *A*.

Hang from the needle a plumb-line which has been rubbed with chalk. By plucking the line and letting it spring back a chalk line may be traced on the board. Repeat the process using the other hole *B*. The intersection of the traces is the centre of gravity.

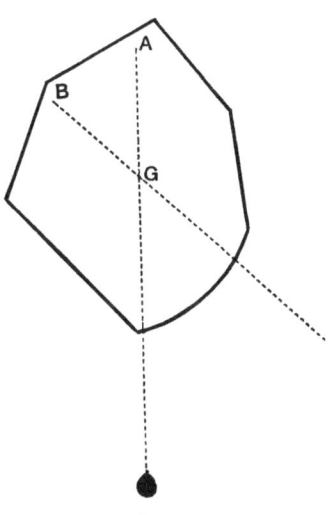

Fig. 17.

19. The centre of gravity can be found by inspection whenever we can discern a point about which the object is symmetrical in all directions. This was the method adopted by Archimedes in his proof of the Principle of the Lever. (§§ 2—3.)

It will now be clear, as observed at the time, that the Principle of the Centre of Gravity is nothing but the Principle of the Lever in its most general form. The rest of his treatise is devoted to finding the centres of gravity of some of the more familiar geometrical figures.

Thus, the c.g. of a straight line is its middle point, for it may be divided into pairs of particles equidistant from the centre on opposite sides.

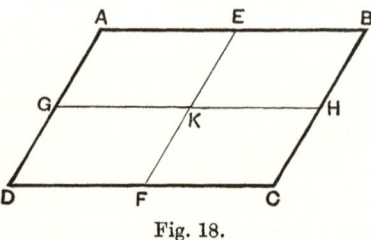

The c.g. of a circle or of a sphere is its centre.

A parallelogram may be divided into strips parallel to one side, AB, each of which is bisected by EF joining the middle points of AB, CD.

Fig. 18.

The c.g. therefore lies in EF. Similarly, it lies in GH. Therefore it is K.

20. To find the centre of gravity of a triangle, ABC.

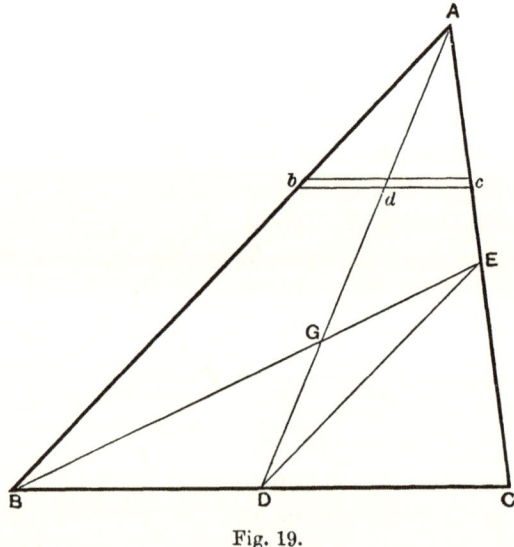

Fig. 19.

Bisect the base in D, and join AD.

Divide the triangle into small strips, such as bdc parallel to the base BDC. Then each strip is bisected, for by similar triangles

$$\frac{bd}{BD} = \frac{Ad}{AD} = \frac{dc}{DC}.$$

But $BD = DC$.   Therefore $bd = dc$.

Hence the C.G. of each strip, and therefore of the whole triangle, lies in AD.

Similarly, it lies in BE, if E is the middle point of AC.

∴ it is the point G.

Join DE.   Then since CB, CA are bisected in D and E, DE is parallel to AB, and we have by similar triangles

$$\frac{GD}{GA} = \frac{DE}{AB} = \frac{DC}{BC} = \frac{1}{2}.$$

The C.G. of the triangle is therefore on the line joining the middle point of the base to the vertex, at one-third of its length from the lower end.

21.   To find the C.G. of any number of weights spaced out along a straight line.

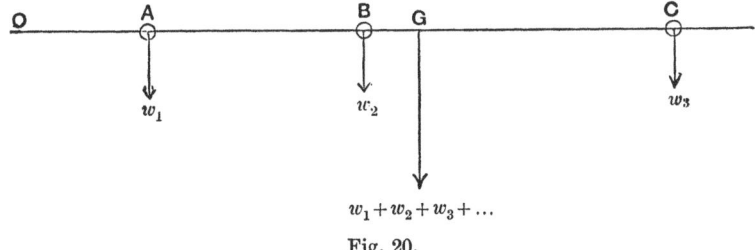

Fig. 20.

If the C.G. were supported by a force equal to the sum of the weights, the whole would remain at rest; and this would not be altered if the rod were produced and fixed at any point, say O.

But the rod would turn about O unless the moments of the separate weights about O were equal to the moment of the supporting force in the opposite direction.

$$\therefore (w_1 + w_2 + \dots) \times OG = w_1 \times OA + w_2 \times OB + \dots$$

$$OG = \frac{w_1 \cdot OA + w_2 \cdot OB + \dots}{w_1 + w_2 + \dots}.$$

Rule.   Hence the distance of the C.G. to the right of the vertical through any point O is found by dividing the sum of the products of each weight by its distance from this vertical by the sum of all the weights.

Very often an object has a line of symmetry, and consists of portions with known centres of gravity spaced out along this line. It is then easy to find its C.G. by the above rule.

Since the object may be turned so as to have any line within it vertical, the same rule will give the distance of the C.G. from any line we choose. The C.G. of a flat body of any shape may thus be fixed by finding its distance from two chosen lines at right angles to each other.

22.　One point of support.

If an object is supported at one point, it will be in equilibrium
Equilibrium　　　　so long as its C.G. is in the vertical line through
under Gravity.　　the point of support. For since its weight then
acts through the point of support, there is no moment tending to
turn it about that point. But the nature of the equilibrium
differs greatly according to the position of the C.G.

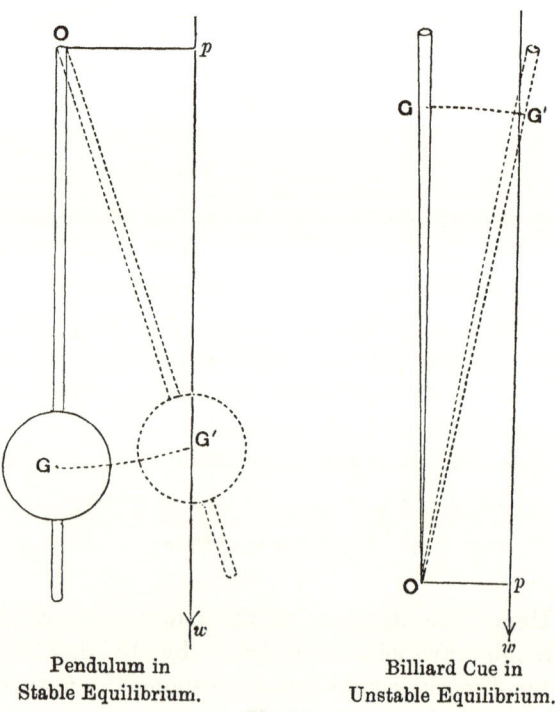

Pendulum in　　　　　　　Billiard Cue in
Stable Equilibrium.　　　　Unstable Equilibrium.

Fig. 21.

(1) Stable Equilibrium. If the C.G. is below the point of support, and the body be accidentally disturbed, the moment of the weight tends to bring it back to its position of rest, and equilibrium is restored. In this case the object is said to be in Stable Equilibrium.

(2) Unstable Equilibrium. If the C.G. is above the point of support, and can descend when the object is disturbed, the moment of the weight tends to turn the object still farther from the position of rest, so that on the occurrence of any accidental disturbance equilibrium is destroyed. This is called Unstable Equilibrium.

(3) Neutral Equilibrium. If the C.G. is *at* the fixed point of support, the object will rest in any position. Such equilibrium is called Neutral Equilibrium. It occurs also when the point of support is moveable so that the C.G. remains always at the same height above it, as when a sphere or cylinder rests upon a horizontal plane.

23. Objects standing upon a base will be in equilibrium so long as the vertical through the C.G. falls within the contour of the base. (If the base has pro-
jecting points and retreating bays, the contour is to be drawn from point to point, and not to follow the inner curve of the bays.) For then the upward pressures from the base can arrange themselves so as to meet and balance the weight acting vertically through the C.G.

Extended Base.

The gesticulations of a person walking along a narrow plank are instinctive efforts to bring back (by shooting out an arm or a leg) the C.G. of the body to the vertical over the line joining the feet. The tight-rope dancer aids himself by a balancing pole heavily weighted at each end. A slight motion of the pole and weights suffices to move the C.G. as much as a violent movement of the limb, and thus awkward and inelegant gyrations are avoided.

24. In bicycle riding the greater part of the balancing depends on this principle, though some help for steadiness is derived from two dynamical considerations to be mentioned later (§§ 70, 77).

Fig. 22.

Let AB be the ground contacts of the front and back wheels, seen from above; G the c.g. of machine and rider. If G moves off the base line AB, say to the right, the rider at once feels that he is falling over on that side. By turning the front wheel towards the side on which he is falling, he brings A to A', while B follows along BA to B', and the base is again beneath the c.g. Hence the rule, so contrary to the beginner's instincts, that the wheel must be turned towards the side on which the rider is falling.

EXAMPLES.

1. Shew that the centre of gravity of a triangle is the same as that of three equal weights placed at its corners.

2. From a body of weight W and centre of gravity G a portion is cut away whose weight is W' and centre of gravity G'. Shew that the centre of gravity of the remainder is G'', on $G'G$ produced, where $G''G = \dfrac{W'}{W - W'}\, G'G$.

3. Weights of 2, 4, 6, 8, 10, 12 lbs. are spaced out along a straight line at equal distances of one foot. Find their centre of gravity.

4. A figure is formed of a square of side a and an isosceles triangle described on one of the sides as base. Find the altitude of the triangle in order that the figure may balance about that side.

5. A sphere of 6 inches radius has a hollow spherical cavity of 2 inches radius, midway between the centre and the surface. Find the distance of the c.g. from the centre.

6. Where must a circular hole 2 inches in diameter be punched out of a circular plate 5 inches in diameter in order that the distance of the c.g. from the centre may be half an inch?

7. The mass of the moon is ·01137 times that of the earth. Taking the earth's radius at 3963 miles and the distance of the moon from the earth's centre at 60·27 radii of the earth, find the c.g. of the earth and moon.

8. Shew that the c.g. of a pyramid on a triangular base is in the line joining the vertex to the c.g. of the base at one-quarter of its length from the c.g. of the base. (Consider slices parallel to the base and proceed as in § 20.)

9. A cylindrical tin can (without lid) 8 inches in diameter and one foot high, is half filled with water. Find the c.g. of the can and the water, if the weight of the can is one-quarter of that of the water.

10. A rod balances about a point one-quarter of its length from one end. If a weight of 2 lbs. is attached to the thin end, the balancing point is shifted 8 inches towards that end ; whereas 8 lbs. must be attached to the thick end to shift it the same amount in the other direction. Find the weight and length of the rod.

CHAPTER IV.

THE BALANCE.

25. ONE of the most important cases of the Lever is the Balance.

In principle it is only a lever with equal arms. If two weights placed at their extremities balance each other, they must be equal.

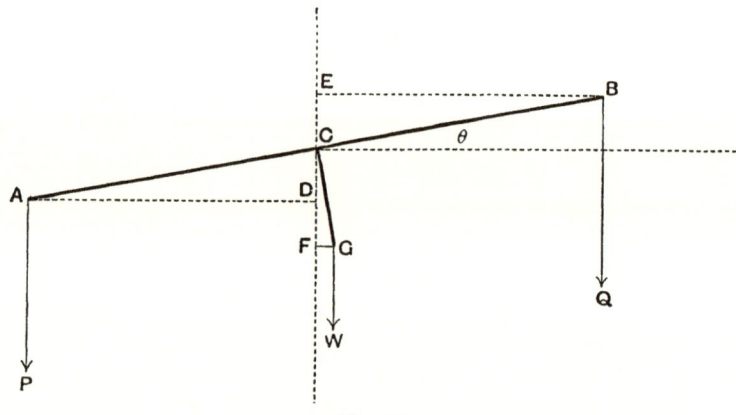

Fig. 23.

Let AB be the beam of the balance, supported at its middle point C. Let $AC = CB = l$; let P and Q be the weights to be compared; W the weight of the beam itself, which will act at its centre of gravity G.

We shall suppose that G is at a distance h below C when the beam is horizontal. Consider what would happen if G were (1) above C, (2) exactly at C. (§ 22.)

If the beam comes to rest at an angle θ with the horizon, we have by the principle of the lever

$$P \times AD = Q \times BE + W \times GF,$$

$$P \cdot l \cos \theta = Q \cdot l \cos \theta + W \cdot h \sin \theta,$$

$$\tan \theta = \frac{(P - Q) \cdot l}{W \cdot h}.$$

If the weights are equal, $\theta = 0$, and the beam is horizontal.

26. A balance is said to have great sensitiveness when a very small difference of weights causes a great deflection. To construct a sensitive balance, we must make l large, and W and h small, *i.e.*

(1) the beam must be long;

(2) the beam must be light;

(3) the centre of gravity of the beam must be very near (but not at) the point of support.

Besides these three requisites the mechanician has also to arrange,

(4) that C shall be exactly on the line AB (for if it is above or below, the sensitiveness will be different for different loads. Ex. 8);

(5) that there shall be as little friction as possible at the points where the beam and the weights are supported (for friction would hinder the free turning of the beam, and perhaps cause the weights not to hang exactly in the vertical through the points of support);

(6) that the time of swing shall not be too great.

The conditions for this last requisite will be understood later (§ 270), but they cannot be satisfied consistently with (1) and (3). Hence a compromise must be effected. Where accuracy is all-important (6) must be given up, and weighing will occupy much time. For rapid, but rough, weighing (1) and (3) are sacrificed.

27. Fig. 24 represents a 16-inch Oertling Balance.

The metal beam is constructed like a girder so as to combine lightness with great rigidity. The point of support is a knife-edge of polished agate projecting downwards at the centre and resting on an agate plane. Two other agate knife-edges project upwards at the ends of the beam, and on these rest agate planes

from which the scale-pans are hung. Above the centre of the
beam may be seen the gravity bob, a small brass weight which
moves up and down on a fine screw.   By means of it the centre of
gravity may be raised or lowered very gradually, and can be
adjusted so as to be sometimes not more than one thousandth of
an inch below the knife-edge supporting the beam. A small vane,
or flag, is also seen, which may be turned to the right or left so as
to correct slight deviations from the horizontal when the beam is
unloaded.

Fig. 24.

A long pointer, attached at right angles to the centre of the
beam, moves over the divisions of an ivory scale at the foot of the
pillar, and should point to zero when the beam is horizontal.

To protect the agate edges from injury while the weights are being changed, and from unnecessary wear and tear, a supporting framework can be raised (by turning the knob A) so as to gently lift the beam off the agate plane on which it rests, and the bars supporting the scale-pans off their knife-edges.

The beam between the central knife-edge and one or both of those at its ends is graduated into ten equal parts, and each of these has again ten divisions. By a lever worked from outside the case a small "rider" (Ω) of platinum wire, weighing one centigram, can be placed on any division of the beam. The case protects the balance from disturbance by air currents during the final stages of the weighing. It rests upon four levelling screws, and is provided with two spirit levels at right angles to each other.

28. With a properly constructed and well adjusted balance

Use of the Balance.

it is for most purposes enough to proceed as follows.

Lower the supports by gently turning the knob till the beam is free, and see whether the pointer rests at zero, or swings to equal distances on each side of it, when the pans are empty.

Raise the supports again, and place the object to be weighed in one scale-pan, and weights in the other. A slight turn of the knob will shew which way the pointer begins to move, and after again supporting the beam, weights must be added or subtracted till, when the beam is set free, the pointer swings slowly backwards and forwards within the limits of the scale. The case may then be closed, and the "rider" placed on the beam, and shifted till the pointer rests at zero, or swings equally on both sides of it. The beam must be brought to rest on the supports every time the rider is to be moved.

The weights in the scale-pan are then recorded before any of them are removed, and allowance made for the rider as follows.

The rider weighs 10 milligrams. If it is placed over the division marked 7, for instance, it will balance a weight of 7 milligrams placed in the other scale, (which hangs at division 10 on the other arm), by the principle of the lever, for

$$10 \times 7 = 7 \times 10.$$

Since each division of the beam is subdivided into tenths and we can estimate tenths of one of the subdivisions, we have the means of reading to hundredths of a milligram.

29. But if such accuracy as this is desired, it is better to proceed by what is known as the Method of Oscillations. We do not wait for the beam to come to rest, but calculate the point at which it will stop in course of time by observing the oscillations of the pointer on the scale.

Suppose the scale has 10 divisions on each side of the zero, and we read the turning points to tenths of a division for three successive swings thus:

Left	Right
	7·3
6·2	
	7·1

The mean of the two readings on the right is 7·2; and the mean of this with the reading on the left is $\dfrac{-6·2+7·2}{2} = ·5$. This is where the pointer would come to rest if left to itself for a quarter of an hour.

It would not be right to take the mean of one observation on each side. For the swings are gradually decreasing, and we ought to compare with the single swing on the left, $-6·2$, such a swing to the right as would have been made at the same stage of their decay. The vibrations decay very regularly, so that the mean between the two swings on the right will represent what a right-hand swing would have been, had it been made at the moment when the pointer was actually swinging 6·2 divisions to the left. We may with advantage observe more than three swings, but there must always be one more swing on one side than on the other.

Using this method we determine:

(1) the position of rest with the pans empty. This may be called the true zero reading;

(2) the reading when weights and rider are adjusted to the nearest whole milligram less than the object;

(3) the reading when the rider is shifted to increase the weight by one milligram.

Thus let the readings be:

(1) Pans unloaded (true zero) ... ... $+ \cdot5,$
(2) With object and weights 36·324 gms. ... $- 4\cdot2,$
(3) „   „   36·325 gms. ... $+ 8\cdot6.$

The addition of 1 mgm. makes a difference of 12·8 divisions. This measures the sensitiveness of the balance for the given load.

To bring the pointer from position (2) to the true zero a difference of 4·7 divisions must be effected, and hence a weight of 4·7/12·8 mgms., i.e. ·37 mgms., must be added.

The true weight is therefore 36·32437 gms.

30. Such perfection has been attained in the construction of balances that a difference of one milligram may be detected in a load of one kilogram, i.e. one part in a million. In a first-class balance the arms may be so nearly equal in length as not to differ by one part in 50,000 ; the knife-edges so keen as to be less than one two-hundred thousandth of an inch wide ; and the centre of gravity may be less than one-thousandth of an inch below the point of support. Such instruments demand great care in handling, and the following precautions should be strictly observed by those who use them.

(1) No one should alter any of the adjustments except those responsible for the care of the instrument.

Precautions.

(2) No change in the object, weights, or position of the rider should be made, nor must the scale-pans or any part of the swinging system be touched except after the beam has been arrested by turning the milled knob.

(3) The knob should be turned gently, and so as to arrest the beam as nearly as possible at the middle of the swing. The great object is to avoid the smallest jerk or jar, as these are likely to injure the agate knife-edges.

(4) The weights must not be touched except with the pliers provided for the purpose.

31. The chief source of error in an accurate weighing, so far as the balance itself is concerned, is a slight difference in the lengths of the arms. This error may be avoided in two ways.

(1)   Borda's method consists in counterpoising the object by weights, small shot, fine sand, or thin paper; and then substituting standard weights for the object till they exactly balance the counterpoise.   It is clear that the arms need not be equal in this method.

(2)   Gauss devised the method of weighing the object in each scale successively.

Let a, b be the lengths of the arms; W_1, W_2 the apparent

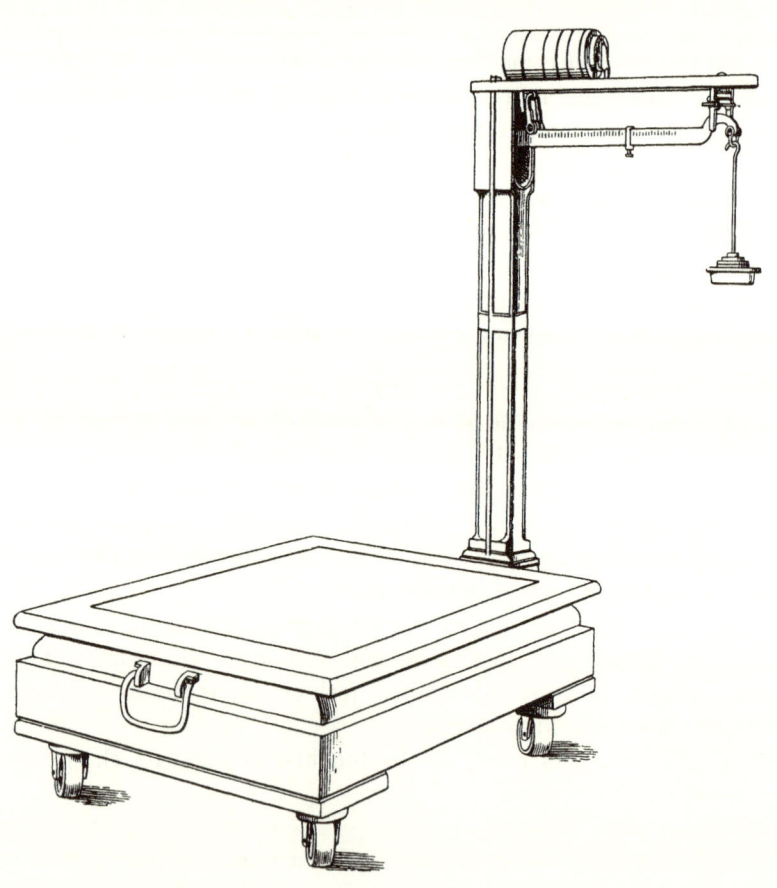

Fig. 25.

weights when the object hangs at a and b respectively; W the real weight. Then by the principle of the lever

$$W \times a = W_1 \times b$$
$$W \times b = W_2 \times a$$
$$W^2 = W_1 . W_2$$

and
$$W = \sqrt{W_1 . W_2}.$$

Since W_1 and W_2 are very nearly equal, it is generally accurate enough to take the arithmetical mean instead of the geometric mean.

32. As an instance of a balance for rough but rapid weighing we may take the common steelyard, to be found at any railway station.

The steelyard.

A platform is attached rigidly to the short end of a balance, very near the fulcrum. Heavy weights can be hung on a hook at the end of the long graduated arm, and a small rider slides on the latter, as in the fine balance just described. It is evident that the divisions can be adjusted so as to read in any units that may be desired.

EXAMPLES.

1. A balance is horizontal when unloaded. But an object weighs 20·4 gms. in one scale and 20·8 gms. in the other. What is the matter with the balance and what is the true weight?

2. An object weighs 12 and 14 gms. respectively in the two scales of a balance. What is the error if its weight is taken as 13 gms.?

3. The pans of a balance are not quite of the same weight, but the arms are equal. Shew that the weight of an object which appears to weigh W_1 and W_2 in the two scales is $\dfrac{W_1 + W_2}{2}$; and that the difference of the weights of the pans is $\dfrac{W_1 - W_2}{2}$.

4. A body whose weight is 12 lbs. appears, in one scale of a balance, to weigh 12 lbs. 6 oz. Find its apparent weight in the other scale.

5. One arm of a balance is 9 inches long and the other 10 inches. Shew that if the seller puts the substance to be weighed as often in one scale as in the other, he loses $\frac{5}{9}$ % on his transactions.

6. In a balance with unequal arms P appears to weigh Q, and Q appears to weigh R ; what does R appear to weigh ?

7. The beam of a balance is 18 inches long, and an object appears to weigh 20·34 gms. in one pan, and 20·87 gms. in the other. How much must the fulcrum be shifted to make the balance true ?

8. In a balance the distance between the knife-edges supporting the scale-pans is $2l$. The central knife-edge is at a perpendicular distance x above the middle point of this line, and the centre of gravity is distant h below the central knife-edge. If weights w_1, w_2 are placed in the pans, and w be the weight of the balance, shew that the beam will come to rest at an angle θ with the horizon, where

$$\tan \theta = \frac{(w_1 - w_2)\, l}{(w_1 + w_2)\, x + w \cdot h}.$$

Hence shew that the *sensitiveness* decreases as the loads increase.

CHAPTER V.

"Wonder en is gheen wonder."

33. THREE of the Mechanical Powers remain to be considered, viz., the Inclined Plane; the Wedge, or Double Inclined Plane; and the Screw, which consists of an inclined plane wrapped round a cylinder, and may be regarded as a Travelling Inclined Plane.

From the time of Archimedes nothing of importance was effected in Mechanical theory for nearly two thousand years, when Simon Stevin of Bruges (1548—1620) established the principle of the Inclined Plane. His discovery constitutes the second step in the historical development of Mechanics. Its importance, and the beautiful ingenuity of the proof, make it worth while to study the proof in his own words.

Stevin not only built upon this foundation the theory of pulleys and the lever, and many propositions of modern Mechanics, but applied his knowledge to practical questions such as, for instance, the design of the machines by which the Dutch fisher-men hauled their boats above high-water mark; the best form of bit for the management of horses (at the request of Maurice of Nassau, Prince of Orange); and the art of fortification. Readers of *Tristram Shandy* will remember that his work was Uncle Toby's constant companion.

34. The problem to be solved was this.

A body resting upon a horizontal plane requires no force to

support it. Let it be attached by a string to some point in the
plane, and let the plane be tilted
till it becomes vertical, so that
the body hangs freely by the
string. The tension of the string
must now be equal to the full
weight of the body. In the intermediate positions the tension will
be something between the weight
of the body and zero. What is
the law connecting the tension
and the weight for any given
slope of the plane? That is the
principle to be discovered.

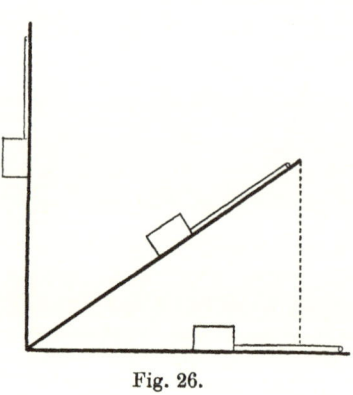

Fig. 26.

35. Here is Stevin's solution, arrayed in all the elaborate
Stevin's Prin- forms of a proposition of Euclid. (*Elements of*
ciple. *Statics*, Book I. Proposition XIX.)

If a triangle has its plane perpendicular to the horizon, and
its base parallel to it; and on each of the two other sides a
spherical mass of equal weight; the power of the left-hand
weight shall be to the power of the right-hand weight as the
right side is to the left side.

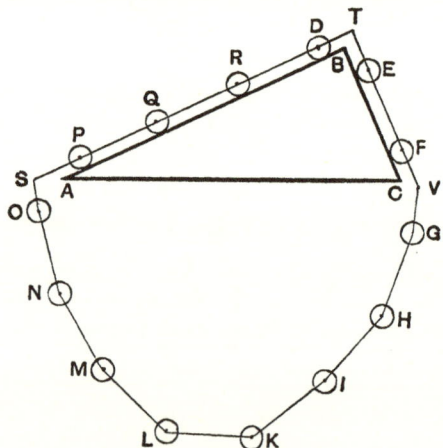

Fig. 27.   (From Stevinus.)

Given. Let ABC be a triangle, having its plane perpendicular to the horizon, and its base AC parallel to the horizon; and let there be on the side AB (which is double of BC) a globe D, and on BC another E, equal in weight and magnitude.

Required. To prove that as the side AB (2) is to the side BC (1), so is the power of the weight E to that of D.

Preparation. Let there be fitted round the triangle a circuit of fourteen globes, equal in weight and size, and equidistant, as $D, E, F, G, H, I, K, L, M, N, O, P, Q, R$, threaded on a cord passing through their centres, so that they can turn on the said centres, and that there may be two globes on the side BC, and four on AB; and thus as line is to line, so the number of globes to the number of globes; let there be three fixed points at S, T, V, over which the cord can slip, and let the two parts above the triangle be parallel to its sides AB, BC; so that the whole can turn freely and without hindrance on the said sides AB, BC.

If the power of the weights D, R, Q, P, be not equal to the power of the two globes E, F, the one side will be more powerful than the other. If it be possible, let the four D, R, Q, P, be more powerful than the two E, F; but the four O, N, M, L, are equal to the four G, H, I, K; wherefore the side of the eight globes D, R, Q, P, O, N, M, L, will be more powerful in consequence of their arrangement than the six, E, F, G, H, I, K, and since the heavier side overcomes the lighter, the eight globes will descend and the other six will rise.

Demonstration.

Let it be so, and let D arrive where O is at present, and so for all the others; viz. let E, F, G, H, arrive where P, Q, R, D are now, and I, K, where E, F are. Nevertheless the circuit of globes will have the same configuration as before, and for the same reason the eight globes will have the advantage in weight and in falling will cause eight others to come into their places, and so this movement will have no end, which is absurd.

The proof will be the same for the other side.

Therefore the part of the circuit, D, R, Q, P, O, N, M, L, will be in equilibrium with the part, E, F, G, H, I, K.

Take away from the two sides the weights which are equal and similarly situated, as are the four globes, O, N, M, L, on the one part, and the four, G, H, I, K, on the other part.

The remaining four, D, R, Q, P, will be, and will remain, in equilibrium with the two, E, F.

Wherefore E will have double the power of D.

As the side BA (2) is to the side BC (1), so is the power of E to the power of D.

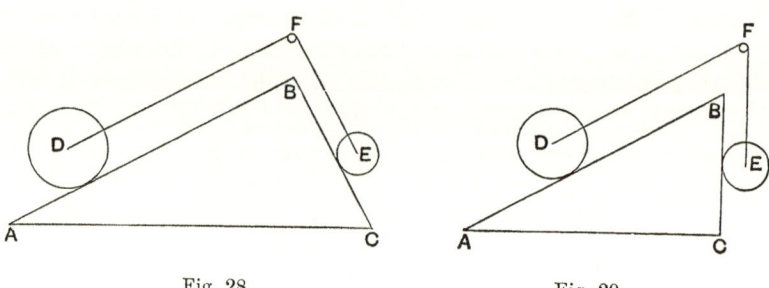

Fig. 28.   Fig. 29.

Corollary 1. Let ABC be a triangle as before, and AB double of BC; and let D, a globe on AB, be double of E on BC. It appears that D, E will be in equilibrium.

Wherefore as AB is to BC, so is the globe D to the globe E.

Corollary 2. Let now one of the sides of the triangle, as BC (which is half of AB), be perpendicular to AC. The globe D, which is double of E, will still be in equilibrium with E. For as the side AB is to BC, so is the globe D to the globe E.

In the last corollary the tension of the string supporting D on the inclined plane is evidently equal to the weight of E hanging freely.

Thus the principle is reached that *the force required to support a body resting on an inclined plane is to the weight of the body as the height of the plane is to its length* (*along the slope*).

36.   Before going farther the student should immediately test
Experimental   this important principle for himself.
Verification.   *Experiment.* With the apparatus figured, or some simpler arrangement, make a series of observations with

different slopes and different weights and verify that in every
case
$$\frac{P}{W} = \frac{\text{Height of Plane}}{\text{Length of Plane}} = \sin \alpha,$$
where α is the inclination of the inclined plane to the horizontal.

Fig. 30.

37.   In all probability the student, especially if familiar with
Criticism of   mathematical demonstrations, will feel a curious
Stevin's Proof.   sudden enlightenment, and intensity of conviction
on first grasping the point of Stevin's proof,—far more than he
will derive from the results of his direct experiments, which will
give only approximate values owing to the effects of friction and
the difficulty of all exact measurements.   How is this?   Do the
truths of science rest after all on *à priori* reasoning, rather than

on observation and experience? The following remarks of Professor Mach on this point are so instructive that I venture to quote them in full; they should be carefully studied.

"Unquestionably in the assumption from which Stevinus starts, that the endless chain does not move, there is contained primarily only a purely instinctive cognition. He feels at once, and we with him, that we have never observed anything like a motion of the kind referred to, that a thing of such a character does not exist. This conviction has so much logical cogency that we accept the conclusion drawn from it respecting the law of equilibrium on the inclined plane without the thought of an objection, although the law if presented as the simple result of experiment, or otherwise put, would appear dubious. We cannot be surprised at this when we reflect that all results of experiment are obscured by adventitious circumstances (as friction, &c.), and that every conjecture as to the conditions which are determinative in a given case is liable to error. That Stevinus ascribes to instinctive knowledge of this sort a higher authority than to simple, manifest, direct observation might excite in us astonishment if we did not ourselves possess the same inclination. The question accordingly forces itself upon us: Whence does this higher authority come? If we remember that scientific demonstration, and scientific criticism generally, can only have sprung from the consciousness of the individual fallibility of investigators, the explanation is not far to seek. We feel clearly that we ourselves have contributed nothing to the creation of instinctive knowledge, that we have added to it nothing arbitrarily, but that it exists in absolute independence of our participation. Our mistrust of our own subjective interpretation of the facts observed, is thus dissipated."

"Stevinus' deduction is one of the rarest fossil indications that we possess in the primitive history of Mechanics, and throws a wonderful light on the process of the formation of science generally, on its rise from instinctive knowledge. We will recall to mind that Archimedes pursued exactly the same tendency as Stevinus, only with much less good fortune. In later times, also, instinctive knowledge is very frequently taken as the starting point of investigations. Every experimenter can daily observe in himself the guidance that instinctive knowledge furnishes him. If he succeed in abstractly formulating what is contained in it, he will as a rule have made an important advance in science."

"Stevinus' procedure is no error. If an error were contained in it, we should all share it. Indeed, it is perfectly certain that the union of the strongest instinct with the greatest power of abstract formulation alone constitutes the great natural enquirer."

(Mach, *The Science of Mechanics*. Translation by Thomas J. McCormack, p. 26.)

And again: "The reasoning of Stevinus impresses us as so highly ingenious, because the result at which he arrives apparently contains more than the assumption from which he starts. If Stevinus had distinctly set forth the entire fact in all its aspects, as Galileo subsequently did, his reasoning would no longer strike us as ingenious; but we should have obtained a much more satisfactory and clear insight into the matter."

We shall see later how Galileo regarded this principle.

EXAMPLES.

1. A train of 200 tons weight rests on an incline of 1 in 80. There is a resistance to motion due to friction equivalent to an opposing force of 16 lbs. per ton weight. What force is required (1) to prevent the train from running down hill, (2) just to set it in motion up hill?

2. To pull a waggon up a certain hill the horse has to exert a force equal to the weight of 480 lbs., one quarter of his effort being employed in overcoming friction. If he zig-zags so as to increase the distance travelled by one-third, what force must he exert, supposing friction to be the same as before?

3. Two inclined planes of equal altitude 4 feet, but bases 3 feet and 5 feet respectively, are placed back to back, and two weights connected by a smooth string are balanced across the top, one on each incline. Compare the weights.

CHAPTER VI.

THE PARALLELOGRAM OF FORCES.

38. IN the case of the Inclined Plane considered by Stevinus there are three forces acting on the sustained weight: (1) its weight pulling vertically downwards, (2) the pull of the string along the slope of the plane, (3) the support of the plane itself. This last force can be nothing but a thrust, or pressure, at right angles to the plane; for the plane is supposed to be smooth, *i.e.* incapable of exerting any sideway reaction on objects in contact with it, of the nature of resistance to slip.

Stevinus not only discovered the relation between the tension and the weight, when it was sustained on the inclined plane; but he perceived that the relation between these three forces must be the same, if they are to balance each other, however they are produced, provided that *the form of the machine's motion, i.e.* the directions of the forces, remains the same.

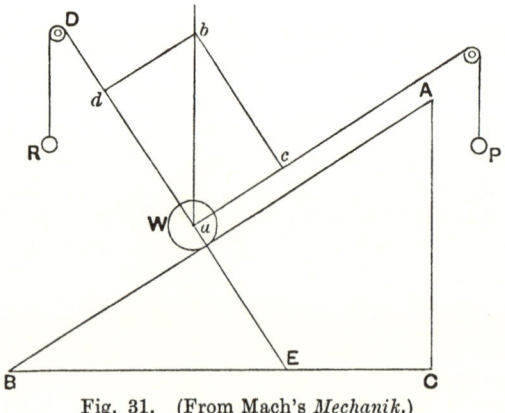

Fig. 31. (From Mach's *Mechanik*.)

For instance, the thrust R might just as well be replaced by the tension of a string perpendicular to the plane, carried over a fixed pulley at D, and supporting a weight R.

The plane may now be removed, and there is left a so-called "Funicular Machine."

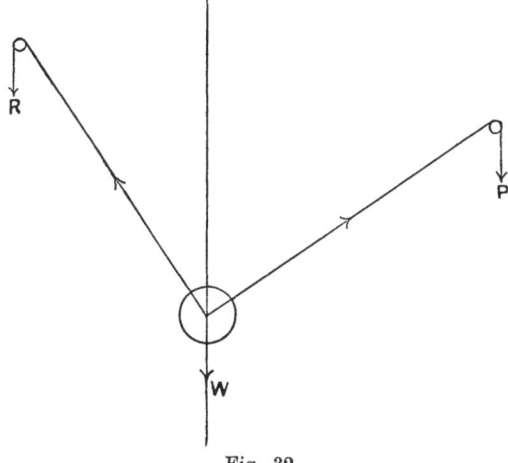

Fig. 32.

Draw the vertical through a, and take any length ab to represent the weight W. From b draw bc perpendicular to AB.

Then by Stevinus' Principle,

$$\frac{P}{W} = \frac{AC}{AB} = \frac{ac}{ab},$$

by similar triangles.

But there is no reason why the function of the two strings should not be interchanged, and R be supposed to support W on a plane DaE, whose reaction is the force P.

In this case

$$\frac{R}{W} = \frac{ad}{ab}.$$

Hence the two forces P and R, which together balance W, are represented by ac and ad, when W is represented by ab in magnitude, but of course in the reversed direction. The line ab must therefore represent the single force which is equivalent to, *i.e.* has the same effect as, P and R acting together, since they just balance W acting downwards.

C.   4

Stevinus was thus led to a special case (when the two forces are at right angles) of a very important proposition, the Parallelogram of Forces, which may be stated as follows:

If two forces acting at a point are represented in magnitude and direction by two straight lines, they are together equivalent to a single force, represented by that diagonal of the parallelogram constructed on the two straight lines which passes through the point.

This proposition, though employed by Stevinus, was first explicitly stated by Newton as a corollary to the Second Law of Motion. It is the starting point of the modern treatment of Statics.

39. *Experiment.* A direct experimental proof of the Parallelogram of Forces is easily arranged.

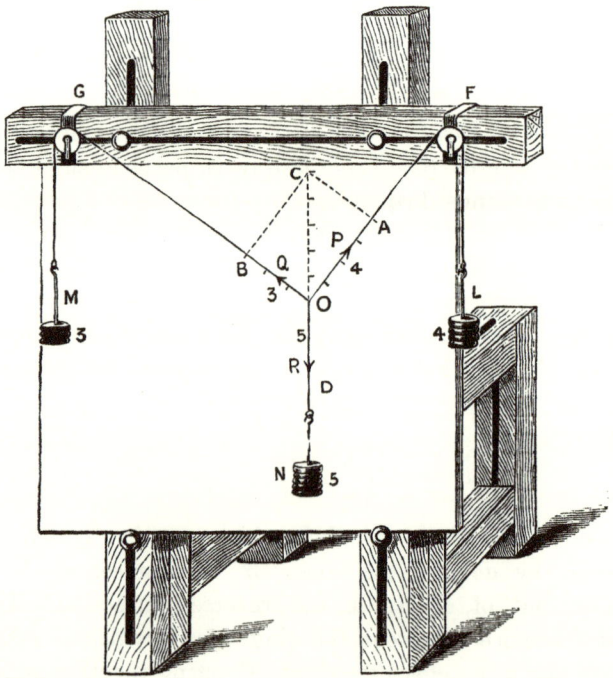

Fig. 33.

Three strings are knotted together, and provided with hooks at the other ends. Two of them pass over light pulleys and have weights attached ; the third hangs vertically, supporting a weight.

The directions of the strings, when equilibrium is attained, can be traced on a drawing board held behind them. The trace of the vertical string is produced upwards, and from any point in it parallels to the other strings are drawn.

It will be found that the sides and vertical diagonal of the parallelogram so constructed are always proportional to the weights hanging from the strings respectively parallel to them.

We shall defer the development of the consequences of this proposition till after it has been deduced from the Laws of Motion.

CHAPTER VII.

THE PRINCIPLE OF VIRTUAL WORK.

40. MECHANICAL problems, and especially the simple machines, may be regarded from another point of view. It was first noted by Stevinus in the case of the pulleys.

When a weight is raised by means of a cord passing over a single fixed pulley, the "Power" must be equal to the weight, and it descends exactly as much as the weight rises.

By employing a single moveable pulley the weight can be raised with half the Power (§ 15); but the cord to which the power is attached must be pulled through twice the height the weight rises.

If pulley-blocks are used, with n strings to the lower block, the force employed need only be one-nth part of the weight (§ 16), but since each of the n strings must be shortened as much as the weight rises, the end of the power string must move n times as far as the weight is lifted.

Stevinus saw that this principle applied to all machines, and embodied it in his phrase:

" Ut spatium agentis ad spatium patientis, sic potentia patientis ad potentiam agentis."

In other words, "What is gained in power is lost in speed."

So that the product of the force exerted and the distance moved through is the same for the Power as for the Weight.

Stevinus shewed how to employ this principle so as to find the relation between the power and the weight for complicated machines. Even when the construction is unknown, it may be used. Imagine (Fig. 33 a) the two handles A and B to be connected by *any* system of mechanism (levers, pulleys, wheel-work, &c.) enclosed in a box.

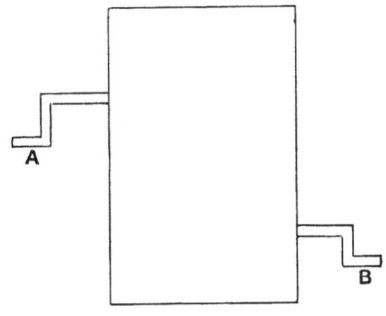

To find what force will be exerted by *B* when a given force is applied to *A*, all that is necessary is to observe how far *B* moves for a given movement of *A*. Then the ratio of the force exerted at *B* to the force applied at *A* is the same as the ratio of the distance moved by *A* to the distance moved by *B*.

Fig. 33 *a*.

41. This was the way in which Galileo regarded the Inclined Plane. He was much occupied with the descent of falling bodies, and was very sure that heavy bodies never rose of their own accord, but settled down under gravity to the lowest place they could reach.

If, then, no motion took place on the Inclined Plane, it must be for the reason that, were the weights to get into motion, there could be no rise or fall of weights *on the whole.*

But when two or more weights are concerned, some rising and some falling, how are we to take into account a large weight rising through a small distance, and perhaps obliquely, as compared with a small weight falling through a large distance? Galileo saw that the essential factors were the weight concerned and the *vertical* distance through which it rose or fell; and that we should carry to our account the product of the two.

Thus (Fig. 31) if *W* is raised from the bottom to the top of the plane *AB*, it rises a vertical height *AC*, while the Power-weight descends a length equal to *AB*. Hence, that there may be no rise or fall of weights on the whole,

$$P \times AB = W \times AC,$$

or

$$\frac{P}{W} = \frac{AC}{AB}.$$

42. Of course Galileo could only have seen that the *vertical* heights were the essential factors from instinctive knowledge based on experience; and the principle could only be finally established by careful comparison with experience according to the canons of logic (*v.* § 118).

"Galileo's conception of the Inclined Plane strikes us as much less ingenious than that of Stevinus, but we recognize it as more natural and more profound *."

When we realize it, we suddenly perceive how the ratio of the forces in the case of the inclined plane fits in with our general experience that heavy bodies settle downwards as far as they can. "The equilibrium equation of the principle may be reduced in every case to the trivial statement, that when nothing can happen nothing does happen *."

"The principle, like every general principle, brings with it, by the insight which it furnishes, disillusionment as well as elucidation. It brings with it disillusionment to the extent that we recognize in it facts which were long before known and even instinctively perceived, our present recognition being simply more distinct and more definite; and elucidation, in that it enables us to see everywhere throughout the most complicated relations the same simple facts *."

43. Torricelli (1608—1647) gave the principle a more general form by employing the notion of the Centre of Gravity.

Let there be a number of weights, P_1, P_2, &c. connected by mechanism; and let their heights above a line of reference Ox be h_1, h_2, &c. Then the height of the centre of gravity is, (§ 21)

$$h = \frac{P_1 h_1 + P_2 h_2 + \dots}{P_1 + P_2 + \dots}.$$

Let the machine work for a moment so that the heights of the weights become h_1', h_2', etc. The new height of the centre of gravity is:

$$h' = \frac{P_1 h_1' + P_2 h_2' + \dots}{P_1 + P_2 + \dots}.$$

The centre of gravity will have fallen a distance

$$h - h' = \frac{P_1 (h_1 - h_1') + P_2 (h_2 - h_2') + \dots}{P_1 + P_2 + \dots}.$$

* Mach, *Mechanik.*

If such a fall is possible, it will certainly take place. If therefore the machine is in equilibrium, it must be because the centre of gravity of the weights attached to it cannot descend, and thus

$$P_1(h_1 - h_1') + P_2(h_2 - h_2') + \ldots = 0,$$

i.e. the sum of the products of the weights into their vertical displacements when motion takes place must be zero.

44. In a letter to Varignon written in 1717 John Bernouilli shewed how to extend the principle to all cases of equilibrium.

Generalization of the Principle.

Let any number of forces act in any directions at any points. Imagine the points to receive any infinitely small displacements compatible with their mechanical connections.

Multiply each force by so much of the displacement of its point of application as takes places along the direction of the force, counting the product positive if the displacement occurs in the same sense as the force, and negative if in the opposite sense.

Then in order that there may be equilibrium the sum of all these products must be equal to zero.

45. We can simplify this statement by introducing the very important term Work.

Work.

In common language any fatiguing exertion is called work. Lifting weights is a simple and familiar example. Consider a number of labourers engaged in carrying bricks, mortar, &c., up vertical ladders to the different floors of a building in course of construction. The amount of work done by any one man depends on two things: (1) the weight of bricks lifted; (2) the vertical height to which they are raised; for it is clear that the man who lifts twice the weight of bricks to the same storey as another man, or the same weight to twice the height, will have done twice as much work.

And it depends on these two things only. It does not depend on the time taken to do it. One man may work steadily, but slowly: another may take frequent intervals for rest and refreshment, and then work furiously. The foreman need not watch them; he can measure the work by the piece, *i.e.* by noting the weight raised and the height to which it is carried.

Nor does it depend on the path by which the bricks are carried. One man may take them up a vertical ladder, by a dead lift through a short distance; another may arrange a series of sloping planks, and arrive with little effort, but after a long walk. The effective result, the work done, is the same if the same weight is raised to the same height.

Work, then, results from two factors,—force exerted, and distance through which it moves its object. Both must be forthcoming for work to be done. Neither is sufficient alone. Great forces are often exerted without doing any work in the scientific sense. The piers of a bridge exert a great upward thrust, but do no work, though they serve a useful end, for they *prevent gravity from doing work,* and so bringing the bridge to the ground.

46. If the object moves, but not along the direction of the force, only so much of the displacement is to be reckoned as takes place along that direction, just as at football it may often be advisable to run with the ball obliquely, but the effective value of the run is estimated by the yards gained in the direct line between goals.

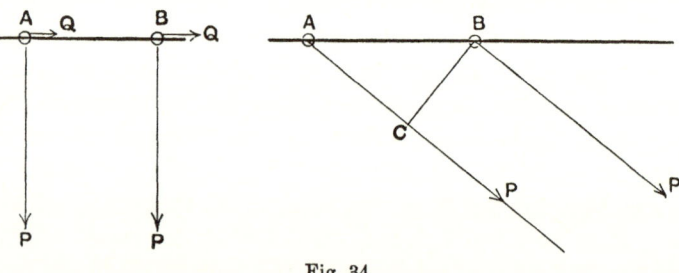

Fig. 34.

Let a curtain ring be pulled with force P by a cord at right angles to the rod. No effect is produced; no work is done. Now let the ring be drawn along the rod by a pull Q, from A to B. The force Q does work, but P does no work in spite of the motion; for it has not effected any advance of the ring in its own direction.

If the same motion is effected by applying the force P obliquely (Fig. 34), P does work. The effective distance through which it has moved the ring is not AB, however, but AC, the projection of AB upon the direction of the force. This projection has the full value AB when P acts along AB, and vanishes when P is at right angles to the motion.

47. In science the term Work is adopted accordingly with the following definition.

Work. A force is said to do work when its point of application is displaced in the direction of the force.

When the displacement is in the opposite direction, work is said to be done against the force, and is counted negative.

The unit of work is the amount of work done by unit force in displacing its point of application through unit length.

If P units of force are acting through a displacement of l units of length, the work done will be $P \times l$ units of work.

The engineer's unit of work is the foot-pound, *i.e.* the work done in lifting one pound through a vertical height of one foot. If the dynamical unit of force is employed, the unit of work is the foot-poundal (§ 124).

On the C.G.S. system the unit of work is the work done when a force of one dyne is exerted through a distance of one centimetre. This unit is called an Erg (§ 124).

48. Returning now to the general principle stated by Bernouilli, we see that the product of each force by the displacement of its point of application along its direction is the work done by the force during the displacement, and is to be counted positive or negative according as the point moves in the direction in which it is urged by the force, or the opposite. And since the work is not really done (for the system remains in equilibrium, and the imaginary displacements are only an artifice to enable us to perceive the relations between the forces required to maintain equilibrium) the word virtual is used to indicate that both displacements and the consequent work done are such as *might* occur, consistently with the structure of the system.

With these conventions the principle of Virtual Work may be stated succinctly as follows:

A system of forces will be in equilibrium, if the total virtual work for any infinitely small displacements consistent with the conditions is zero.

49. The displacements are to be taken infinitely small, for if a finite motion is allowed, the system may pass over into some other configuration, where different conditions of equilibrium may prevail.

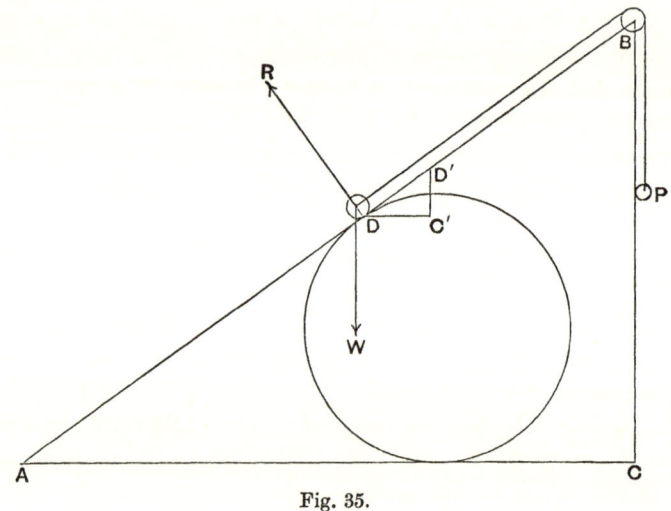

Fig. 35.

This is not always the case. Thus in Fig. 35, the relation between the weights P and W will be the same so long as W remains at D, whether it be supported by the sphere or the plane which touches the sphere at D.

(1) Let it be on the plane. Suppose the weight P to descend till W arrives at D'. Then the vertical rise of W, $C'D'$, is to the vertical fall of P, DD', as BC to AB, *i.e.*,

$$C'D' = DD' \sin A,$$

whether DD' be small or great. The equation of virtual work,

$$W \times DD' \sin A - P \times DD' = 0,$$

i.e.
$$\frac{P}{W} = \sin A = \frac{BC}{AB},$$

holds good equally well for an infinitesimal displacement, or for the whole length of the plane.

(2) But if the weight is resting on the sphere, it is only for an infinitesimal displacement (*i.e.* for so long as the sphere may be taken to coincide with its tangent) that we get $P/W = \sin A$. Farther along the sphere the inclination of the tangent and the ratio P/W are different, so that if we want the conditions of equilibrium at D, we must restrict ourselves to an infinitely small displacement.

We shall now apply this principle to a few cases that are not so easy of solution by other means.

50.   This is only a case of the Inclined Plane, as may be seen by cutting out a right-angled triangle in paper
The Screw.   and wrapping it round a ruler.

The Power is, however, applied parallel to the base of the plane, as in the Wedge; and the plane is made to slide under the weight so as to raise it.

The distance from thread to thread of the screw, measured parallel to the axis, *i.e.* the distance through which the screw

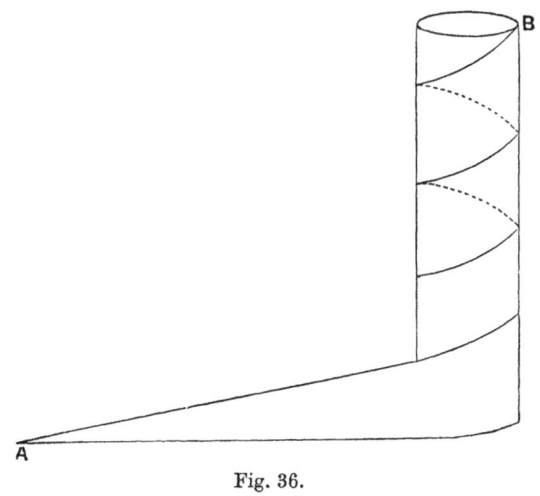

Fig. 36.

advances for one turn of the head or lever arm, is called the Pitch of the screw.

Let the pitch of the screw be p, and the length of the lever-arm l.

As in the inclined plane, the form of the motion, and hence the law of equilibrium, remains unchanged however far the screw be turned. We need not therefore restrict ourselves to an infinitely small motion. Let us suppose the screw to make one complete turn.

Then, apart from friction, the virtual work consists of $P \times 2\pi l$ done by the Power; and $W \times p$ done against the Weight. Thus for equilibrium,

$$P \times 2\pi l - W \times p = 0,$$

and
$$\frac{P}{W} = \frac{p}{2\pi l} = \frac{\text{pitch}}{\text{circumference of power-circle}}.$$

51. Weston's Differential Pulley consists of an upper block with two grooves of slightly different radii, R, r, connected by an endless chain, as in the figure, with a single moveable pulley. The grooves of the upper block contain notches or teeth which fit into the links of the chain so that it cannot slip.

The Differential Pulley.

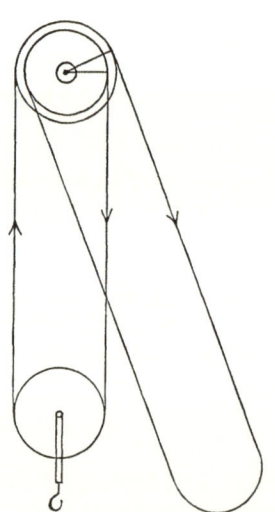

Fig. 37.

Here again we may take the displacement large if we wish. Let the upper block make one revolution. The virtual work done by the Power is $P \times 2\pi R$.

Meanwhile the weight is raised by half the difference between the length of chain wound up on the large groove of the upper block, and that which is unwound from the small one. Hence the virtual work done against the weight is

$$W \times \frac{2\pi R - 2\pi r}{2}.$$

For equilibrium :

$$P \times 2\pi R - W \times \frac{2\pi (R - r)}{2} = 0,$$

$$\frac{P}{W} = \frac{R - r}{2R}.$$

52. A common form of balance for weighing letters is that

Roberval's
Balance.

devised by Roberval.

The scales are attached to the two vertical

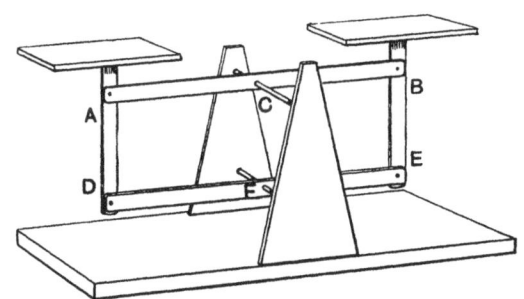

Fig. 38.　Roberval's Balance.

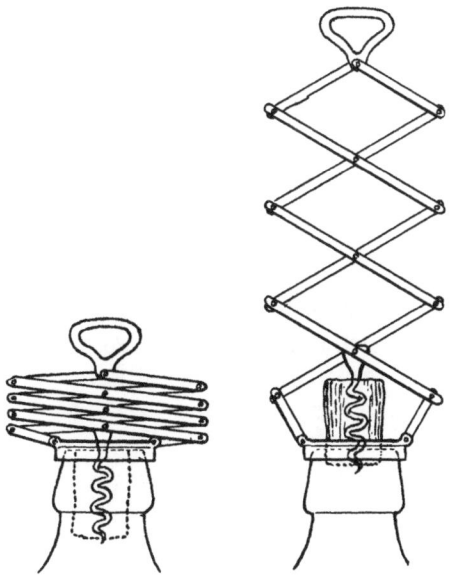

Fig. 39.

sides of a jointed parallelogram, the other two sides turning about pins at their centres.

If the balance moves, one scale descends exactly as much as the other rises. Hence it does not matter where the weights are placed on the scales. If they are equal, their virtual works are equal and opposite, and there will be equilibrium.

53. In the form of corkscrew figured on the preceding page
Lazy Tongs. the handle moves four times as far as the head of the screw. Hence the pull exerted on the screw is four times as great as that applied to the handle.

EXAMPLES.

(Many problems may be solved very easily by the principle of Virtual Work when the Differential Calculus is employed to find the small imaginary displacements of the parts of the system. For example:

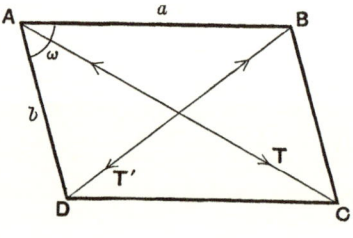

A hinged parallelogram of sides a, b has its opposite corners joined by strings screwed up to tensions T and T''. Find one angle of the parallelogram.

Suppose the parallelogram distorted so that the angle ω is slightly increased to $\omega + d\omega$.

The reactions at each hinge are equal and opposite for the two rods meeting there, so that their virtual work vanishes.

The sum of the v.w.'s for T and T' is

$$T \times d(AC) + T' \times d(BD) = 0.$$

Now
$$AC^2 = a^2 + b^2 + 2ab \cos \omega,$$

$$BD^2 = a^2 + b^2 - 2ab \cos \omega.$$

$$\therefore AC . d(AC) = -ab \sin \omega . d\omega,$$

$$BD . d(BD) = +ab \sin \omega . d\omega.$$

$$\therefore \frac{T}{T'} = -\frac{d(BD)}{d(AC)} = \frac{AC}{BD}.$$

$$\therefore \frac{T^2}{T'^2} = \frac{a^2 + b^2 + 2ab \cos \omega}{a^2 + b^2 - 2ab \cos \omega}$$

whence
$$\cos \omega = \frac{(a^2 + b^2)(T^2 - T'^2)}{2ab(T^2 + T'^2)} .)$$

As examples which may be solved without the Calculus take the following:

1. A light wire is stretched over two smooth pulleys at a distance of 10 feet from each other in the same horizontal line, and has 112 lbs. hung at each end. What weight hung at the middle of the wire will cause it to sag one inch?

2. An elastic ring of natural length l and weight W is laid over a smooth vertical circular cone of angle $2a$. The tension of the ring when stretched to a length l' is given by $T = T_0 \dfrac{l' - l}{l}$. At what depth below the vertex will the ring rest?

54. Let any forces P, Q, R, &c., act at points A, B, C, &c.,

Lagrange's proof of the principle of virtual work. and suppose them to be applied as follows. Let there be pulleys at A, B, C, with other fixed pulleys in the proper directions at A', B', C'. Attach a string at A', and carry it P times back and forth to A; then round B', and Q times back and forth to B,

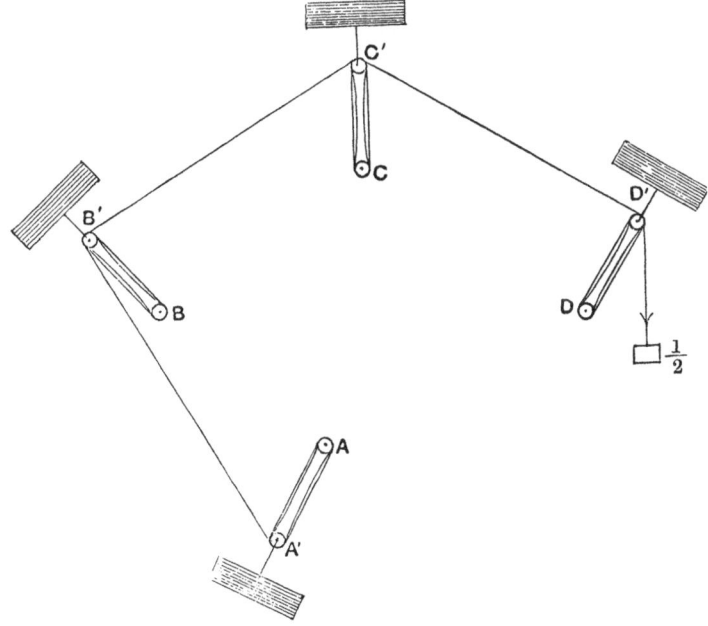

Fig. 40.

and so on for all the forces. Finally let it hang from the last pulley, and attach a weight equal to half a unit. Then since the tension is the same throughout, if the pulleys be smooth, and $2P$, $2Q$, $2R$, &c. strings run to A, B, C, respectively, the forces P, Q, R, will be applied at those points.

Now if among the possible mutual displacements of the points A, B, C, &c., there be any which would on the whole allow the half-unit weight to descend, then it certainly will descend, and work will be done. But if the weight remained at the same level, or had to rise, whatever combination of small movements were given to A, B, C, then motion would not ensue.

Suppose that the result of any such movements of A, B, C, were to shorten the strings between AA' by an amount a, those between BB' by b, and so on. Then for equilibrium the total shortening, *i.e.*, the amount by which the last weight would descend, must be zero, or less than zero.

But the shortening of the $2P$ strings at A is $2Pa$, and so for the others. Thus

$$2Pa + 2Qb + 2Rc + \ldots \gtreqless 0,$$

or
$$Pa + Qb + \ldots \gtreqless 0.$$

The sum of the virtual works is therefore zero or negative.

55. Lagrange's ingenious idea makes it easier for us to understand the principle, for it enables us to fix our attention on the motion of one weight instead of many. But it is not a proof that the possibility or impossibility of doing work is decisive of equilibrium. That principle is involved in each of the pulleys he employs, as much as in the more complicated system. It can only be derived from experience.

56. Lagrange's arrangement also helps us to study the system of bodies A, B, C, acted on by forces P, Q, R, regarded as a machine for doing work. Let the hanging weight carry a pencil pressing against a sheet of paper, carried past it horizontally,—wound, for instance, on a drum with vertical axis. Then if the system be allowed to move (consistently with its mechanical connections) the depth of the hanging weight below its original position will be an indicator of the work done by the system in reaching any other configuration. The pencil will record this depth in a curve, as in Figure **41**.

It was pointed out by Maupertuis in 1740 that when the system arrives at a position of equilibrium, the work done is in general a maximum or a minimum. The weight is at a turning point of the curve.

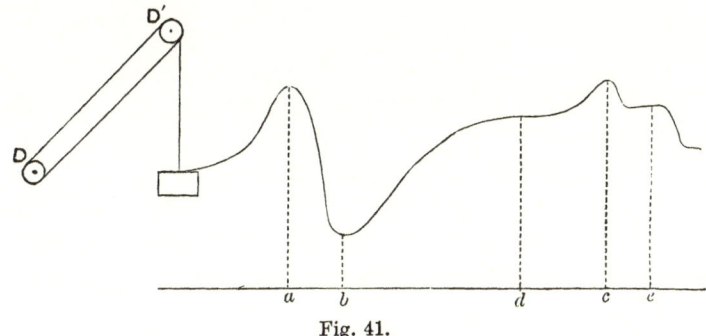

Fig. 41.

When its height is a maximum, as at *a, c*, the system can do work (the weight can descend), if disturbed from the position of equilibrium on either side of it. But when the height is a minimum, as at *b*, the system can only do work by returning to the position of equilibrium if disturbed from it.

Stable equilibrium therefore corresponds to a maximum of work done by the system, unstable equilibrium to a minimum.

If the curve remains horizontal for any finite distance, as at *d, e*, equilibrium exists for all the corresponding positions, with no tendency to pass from one to another. The equilibrium is then neutral equilibrium, as when a sphere rests on a horizontal plane.

CHAPTER VIII.

REVIEW OF THE PRINCIPLES OF STATICS.

57. IN the historical development of Statics the different investigators have adopted different tests for the existence of equilibrium.

Archimedes fixed his attention on the weights and their distances from a fulcrum, and arrived at the principle of the Lever.

Stevinus divined the principle of the Inclined Plane, and referred equilibrium to a relation between the forces and their directions, more fully expressed in the Parallelogram of Forces.

Galileo saw that equilibrium was determined by the weights and their vertical descent towards the earth, and so reached the principle of Work.

Each of these principles is an expression of our experience from one point of view or another. As such they are equally valid. Their authority is coequal, and each is sufficient in itself as a foundation for the science of Statics. We may develope it from any one, or employ them all. Which of them shall be selected is a matter of convenience, or of historical accident.

58. As we might expect, they are mutually deducible.

The Parallelogram of Forces has already been deduced from the Inclined Plane (§ 38), at all events for the case when the forces are at right angles.

Galileo deduced the Inclined Plane from the Lever.

He points out that the ratio of P to W depends on the form of the motion, *i.e.* that W should move along aB, while P descends vertically. It is a matter of indifference whether W is compelled to do this because it rests on a plane AB, or for some other reason, as, for instance, that it should be attached by a bar aO to a fixed

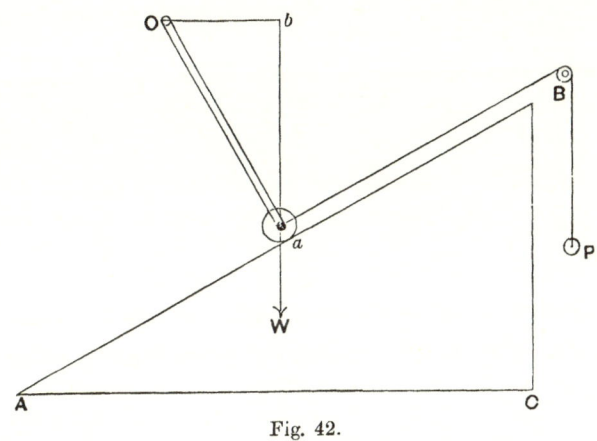

Fig. 42.

pivot at O. It would still have to begin to move along aB, at right angles to Oa.

But in this case by Leonardo's form of the principle of the Lever (§ 11),

$$P \times Oa = W \times Ob,$$

and

$$\frac{P}{W} = \frac{Ob}{Oa} = \frac{BC}{AB}.$$

We shall deduce the Principle of the Lever from the Parallelogram of Forces later (§ 174).

The Principle of the Inclined Plane has been deduced from that of Work (§ 40).

The mutual relation of the Principle of the Lever and that of Work is easily seen.

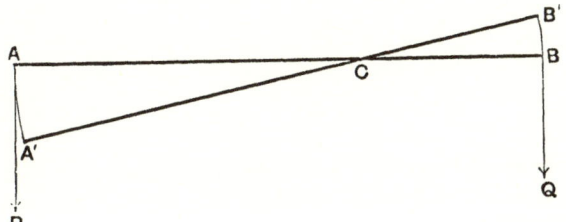

Fig. 43.

Let the Lever ACB receive an infinitely small displacement to a new position $A'CB'$.

The virtual work done by $P = P \times AA'$.

The virtual work done against $W = W \times BB'$.

Assuming the Principle of Virtual Work, we have

$$P \times AA' - W \times BB' = 0.$$

$$\therefore \quad \frac{P}{W} = \frac{BB'}{AA'} = \frac{BC}{AC}.$$

But this is the Principle of the Lever. The converse is obviously true.

59. It is interesting to trace in this manner the connection between the various principles, but it does not increase the authority of any one of them to deduce it from another. Having followed the actual historical order in which they were arrived at, we are at liberty to make any one, or all of them, the starting point of further developments. The modern, and on the whole the most convenient, practice is to deduce the Parallelogram of Forces from Newton's Laws of Motion, themselves but another expression of experience. The other principles, and the whole science of Statics, can then be built on this principle. Statics thus becomes a special case of Dynamics, when the forces concerned happen to be in equilibrium. This is the course we shall now adopt, leaving the further development of Statics till we have traced the discovery of the fundamental principles of Dynamics.

Il Divino Galileo.

Nella Calcografia Giorgi Livorno

CHAPTER IX.

GALILEO AND THE BEGINNINGS OF DYNAMICS.

60. In 1638, when Stevin had already cleared up so much of
The Problem of   Statics, no progress had been made with that part
Falling Bodies.   of the subject, now called Dynamics, which deals
with Motion. The first problem to be considered was, naturally,
the familiar case of the fall of heavy bodies to the earth. Its
solution was the achievement of Galileo, who in the course of his
researches was led to the discovery of several principles of general
importance in Mechanics.

It is not easy at the present day to realize the difficulties
Galileo had to encounter. Let us try to strip ourselves of what is
now common knowledge, and see what were the views held in his
day, with all the authority of two thousand years' acceptance
backed by the great name of Aristotle.

The fall of heavy bodies (and the rise of light bodies which
often accompanied it) was accounted for by assuming that " every
body sought its natural place," and that the place of heavy bodies
was below, that of light bodies above.

Thus in the Elzevir edition of Stevin, Leyden 1634, the editor,
Albert Girard, speaks of " Tant de millions de matières, qui sont
disposées chacunes en leurs lieux," and gives a general definition
of gravity.

" Pesanteur est la force qu'une matière démonstre à son
obstacle, pour retourner en son lieu."

" Ce que je démonstreray, et soustiendray en temps et lieu, à
ceux qui ne le pourront pas comprendre."

When movements were observed in which heavy bodies rose
and light ones fell for a time, such motions were distinguished

from "natural" motions by the term "violent." It was believed that heavy bodies fell more quickly than light ones.

It will be seen that such ideas were too vague to serve as starting points for progress. They were guesses at the reason why bodies fell; attempts to find a cause for their motion.

61. It was already, as Mach has pointed out, a proof of genius that Galileo could so far shake himself free from the prevailing notions of his time, as to take up the modern point of view, and ask himself first *how* bodies fell. That is to say, he began by investigating the facts, and tried to discover the rule or law according to which the fall took place.

Now a falling body starts from rest and passes over a certain distance in a certain time with a speed which a very slight observation shews to be rapidly increasing. The questions which Galileo thought important were such as these: How is the speed acquired by the body in its fall related to the distance it has fallen? Or to the time of fall?

Here again he was met by difficulties, in the lack of experimental means for measuring times and speeds. The mechanical clocks of his day were useless except for considerable lengths of time, and could not be relied on for measuring a few seconds or fractions of a second. One is at a loss to know whether to admire more the ingenuity with which he overcame the experimental difficulties, or his philosophical insight as to the real points to be investigated.

In his treatise, *Discorsi e Dimonstrationi Matematici*, he begins by a guess which seems natural enough at first sight, that the speed acquired will be proportional to the distance the body has fallen from rest. But before putting this to the test of experiment, he examines the hypothesis, and convinces himself that such a rule of motion involves a contradiction, is in fact inconsistent with itself.

62. The next idea that occurs to him is that perhaps the speed will be proportional to the time of fall. Finding no contradiction in this, he proceeds to test it experimentally.

Since it was next to impossible, with the means at his disposal, to measure the speed acquired by a body even in a short fall, he calculates the distance that a body ought to fall through, on the hypothesis that the speed acquired was at each instant proportional to the time elapsed from the start. His proof is a good instance of the use of graphic methods.

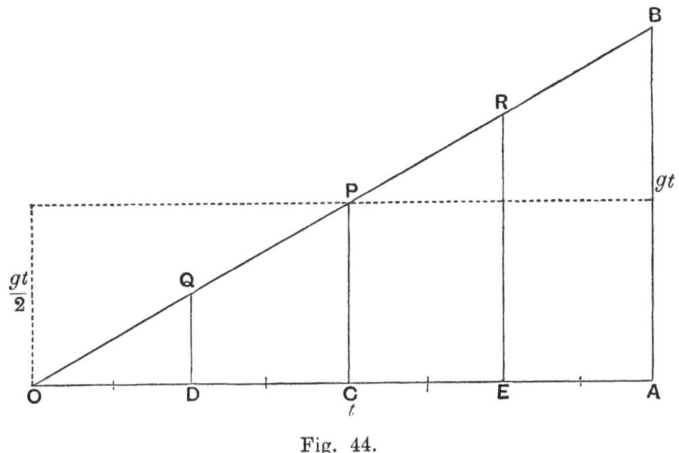

Fig. 44.

63. Let us represent the time elapsed by a straight line OA, which may be divided up so as to represent the different intervals of which the whole time is composed. At each point, such as D, erect a perpendicular whose length shall be proportional to the speed acquired at the moment D. Then, since the speed increases proportionally to the time, the ends of these perpendiculars will lie along the straight line OB.

Let CP be the speed at the middle moment. It is clearly half AB, the final speed.

Consider two points, D, E equidistant from C. The speeds DQ, ER will be the one as much less, as the other is greater than CP, the speed at the middle moment. So that if we compare the real motion with that of a body which should start with the speed CP and maintain it unchanged throughout, we see that any loss of distance travelled, owing to the smaller speed at D, will be exactly compensated for by the greater speed at the corresponding point E. The distances fallen through by the two bodies will therefore be the same in the end.

As we do not know the speed acquired by a falling body during a fall of one second, let us call it g feet per second. Then if the idea that the speed is proportional to the time of fall be correct, the speed at the end of t seconds will be gt. The speed at the middle moment will be $gt/2$, and the distance fallen by a body moving with this speed unchanged for the whole t seconds, will be

$$\frac{gt}{2} \times t = \frac{gt^2}{2}.$$

We could thus make a table for different numbers of seconds, as follows:

Time of Fall	Speed acquired	Space fallen
1 second	g	$g/2 \times 1 = g/2$
2	$2g$	$g/2 \times 2^2 = 2g$
3	$3g$	$g/2 \times 3^2 = 9g/2$
....	....	....
t	gt	$g/2 \times t^2 = gt^2/2$

64. Galileo's Method.

To avoid the difficulties introduced by the great speed acquired
Experimental   by a body falling freely even for one or two
Verification.   seconds, Galileo *assumed* that a ball rolling down an inclined plane in a groove would follow the same kind of rule as a freely falling body, but with diminished speed.

Marking the groove at different distances from the top, he proceeded to measure the times occupied by the ball in reaching the various marks, and verified that the distances travelled really increased as the squares of the times.

For the measurement of the times Galileo made an ingenious modification of the water-clock of Archimedes, which had not been hitherto applied to the measurement of small times. The speed at which water flows out of a hole in the bottom of a vessel depends on the height of water standing above the hole. Galileo took a broad vessel of large area, and hence the level was not appreciably altered during one of his experiments. At the moment when the ball was released he removed his finger from the hole, allowing the water to flow into a vessel which was placed on a balance. When the ball reached a mark, the hole was closed

by the finger again, and the time elapsed could be measured by
weighing the water which had escaped.

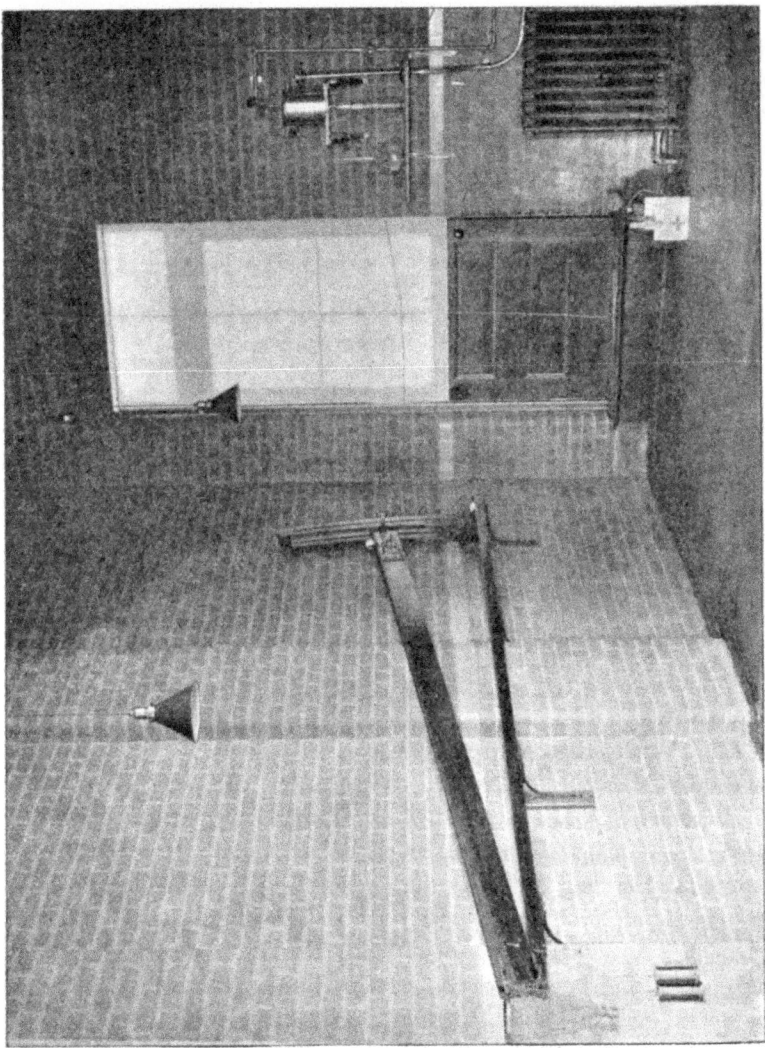

Fig. 45.

Figure 45 represents a modern version of Galileo's apparatus.
It is found that the squares of the times required to reach the
different marks are proportional to the distances of the marks

from the starting point. The student should make experiments
with different slopes of the plane, and plot the results on squared
paper, laying out the distances along a horizontal line, and the
times in the vertical direction. The curve so obtained will be a
parabola. If we choose the vertical ordinates to represent the
squares of the times, instead of the times themselves, we obtain a
straight line.

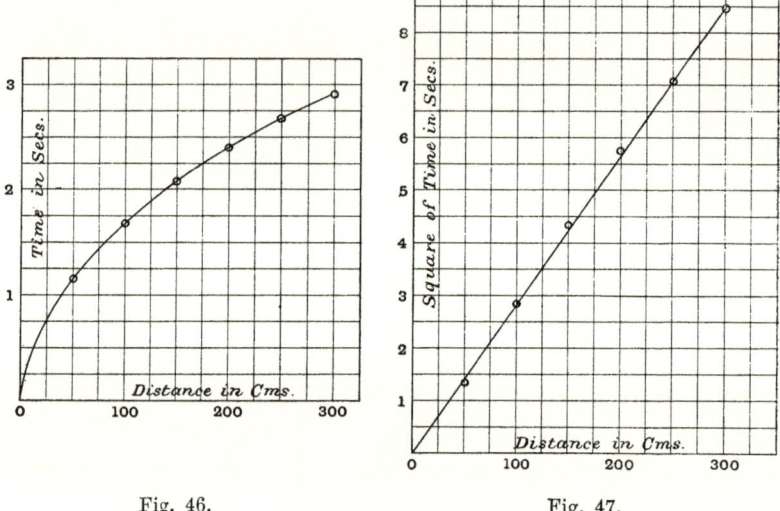

Fig. 46.　　　　　　　　　　　　Fig. 47.

65. With modern apparatus it is easy to verify the same law
for bodies falling freely.

Experiment. Let a plate of smoked glass be suspended as in
the figure, so that when the thread is burnt away, it will fall past
a horizontal tuning fork to which a bristle is attached. The fork,
when sounded, makes a definite number of vibrations every second,
say 100. The result will be a trace on the smoked glass consisting of a number of waves growing longer and longer, since, as
the plate gathers speed, a greater and greater distance will be
travelled in each hundredth of a second.

By measuring the length of a wave at any distance from the
starting point, we can find the speed at which the plate was
moving after falling through that distance. Make several such

measurements. Since the distance fallen is proportional to the square of the time, and the speed proportional to the time, the

Fig. 48.

squares of the speed will vary as the distances fallen through. Plot your results on squared paper, and see if this is the case.

Ideas, of great importance in Mechanics, suggested to Galileo by his experiments.

66. In default of means for studying directly the motion of a body falling freely, Galileo asked himself how the motion of a body sliding down an inclined plane, would be related to that of a body falling freely.

1. The connection between Speed acquired and Vertical Fall.

He concludes that:

The speed attained on an inclined plane must be the same as that attained in falling freely through the same vertical height.

At first sight this seems a startling assumption. But consider what would be the consequence if it were not true.

Let a body slide from *A* to *B*, and then be reflected up the equally inclined plane *BC* (Fig. 49). Galileo feels that it will exactly reverse its motion, losing speed precisely as it gained it along AB, and coming to rest at C, at the height from which it started.

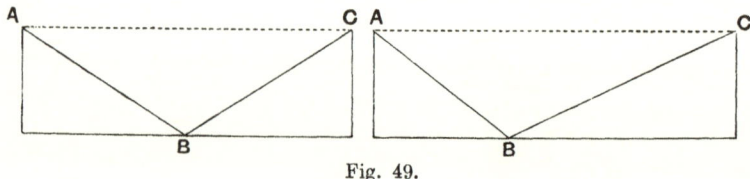

Fig. 49.

Let it next be reflected up a plane of less inclination, BC'. It must still reach the same height. For if it went farther, it would have risen, without external aid, beyond its original height; and by arranging a series of steep and gentle slopes alternately we could make a body, starting from the top of the first slope, raise itself unaided to any height we choose. But this we feel to be contrary to all our experience. We have never met with anything like it. If on the other hand it failed to reach C', we should only have to start the body from C' and it would rise above A, and the same contradiction of experience would arise.

It must therefore rise to C', neither more nor less. Hence the speed attained must be the same for the same vertical fall, whatever the slope of the plane on which it takes place.

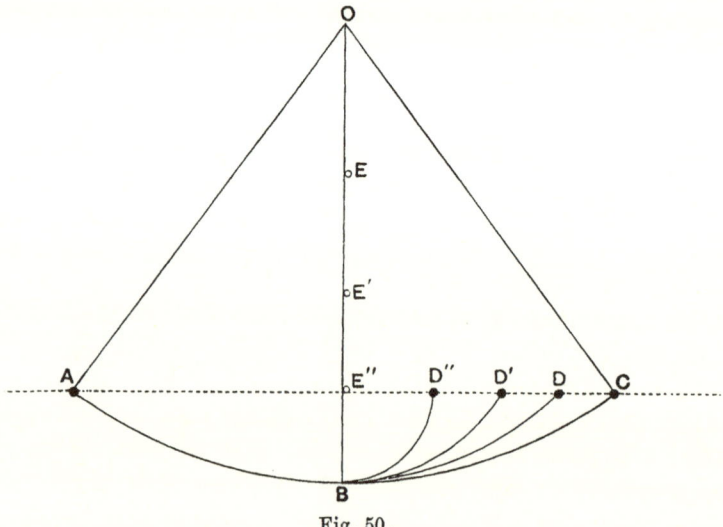

Fig. 50.

67. It is characteristic of Galileo's modern spirit that he at once proceeds to test this conclusion by an experiment, and that a very beautiful one.

Experiment. Hang a bullet by a thread from a nail O, and drawing it aside to A release it. It describes the curve AB, which Galileo perceives may be regarded as a series of very short inclined planes of different slopes, so that his law should hold for the curve as well as for the plane.

It then reverses its motion, arriving at C, in the same horizontal level with A.

Drive a nail in at E, and repeat the experiment. After passing the vertical the bullet will describe a circle of shorter radius BD. But it will rise to exactly the same level, whatever circle we make it describe.

68. Galileo's assumption no longer strikes us as strange when we realize that it involves only the perception that bodies cannot, unaided, raise themselves to a higher level above the earth, a fact which we feel instinctively to agree with all our experience.

With the aid of this assumption, since he could already determine the space fallen through in a given time on an inclined plane, he was able to form a notion of the velocity acquired by a body falling freely. Let us however perform the experiment directly.

69. *Experiment.* Apparatus required:—A pendulum ticking seconds loudly ; or a metronome set to beat seconds or half seconds. A ball which may be dropped from a height of 16·1 feet.

If the ball be released by hand precisely at any tick of the pendulum, it will be heard to strike the floor at the next tick. (The result will be more accurate if the ball be of iron, suspended from an electromagnet whose circuit is broken by the pendulum itself as it makes the first tick. *Experimental Mechanics.* Sir R. Ball.)

Since the ball covers 16·1 feet in one second, its average speed must be 16·1 feet per second. But this (§ 63) is one-half the final speed. Therefore in one second it has acquired a speed of 32·2 feet per second.

70.   We must now attend to an inference which Galileo draws
2. The First Law from his experiments, as it were incidentally, and
of Motion. probably without seeing its full importance.   He
was most likely led to it by the *Principle of Continuity*.   In
geometry a doubtful conclusion may often be tested by trying
whether it is true in an extreme case.   The great investigators
have this principle continually in mind.   "Natura non agit per
saltum."   It is not likely that one law should hold good in one
case, and quite a different one in some other case only slightly
different from it.   Apply this to the case of the body falling down
the inclined plane in figure 49.   What will happen if the second
plane be made more and more oblique ?   It is clear the body
will have to travel farther and farther before it reaches the level
AC', and if the second plane be made ultimately horizontal, the
body will never reach the level, however far it may travel.   But
in this case it would go on for ever !   At the same time we see that
no part of the weight is employed in stopping it.   Hence *a motion
once started, will continue indefinitely if nothing interferes to stop it.*

Now this is an entirely new point of view, not only contrary
to the current ideas of Galileo's time, but surprising to many
uninstructed people even at the present day.   Since all the movements we observe in practice are shortly brought to rest by
various frictions and resistances, it is quite natural to imagine, as
the unobservant do to this day, that every motion requires some
cause or force to maintain it, and ceases when the force is with-
drawn.   Logic will not help us to settle the question.   For
against the principle, "the effect of a cause persists," we may set
another, such as "cessante causa cessat effectus."   Nothing but
experiment can decide whether in this case the "effect" of the
force is the change of place, or the speed acquired.   Now Galileo
perceives from his experiments that the "effect" of the weight of
a body is to produce a change in its *speed**.

We shall see what an advance this was, if we compare the old
idea that "bodies sought their place," that of heavy bodies being
below, with Galileo's notion.   When a stone is thrown upwards,
the first principle seems to be contradicted, for the heavy stone
rises.   But Galileo sees that its weight is changing its speed, as

* See Mach, *Mechanics*, pp. 140—143.

much during the rise as during the fall, reducing it by 32·2 feet per second in every second, till at last it comes to rest, and then begins the descent.

71. The notion of speed as applied to bodies moving uniformly must have been as familiar in Galileo's day as in ours. If a man walks steadily for three hours and finds he has travelled twelve miles, he estimates that he has been walking at the rate of four miles an hour. We may watch him for one hour and find that he goes four miles. Or we may time him for any distance and make the calculation that in one hour he will travel four miles. The relation between t, the number of hours, s, the miles travelled, and v, the speed in miles per hour, is evidently

3. Variable Velocity.

$$v = \frac{s}{t}.$$

But this notion no longer serves us in the case of the falling stone, which changes its speed from moment to moment.

Observe, however, that the rule

$$v = \frac{s}{t}$$

in no way depends on the particular distance we choose to measure. Provided we have the means of determining very short times and distances we may calculate the speed as well from noting the distance travelled in the millionth part of a second, as from that travelled in a day.

Precisely this method is employed in finding the muzzle velocity of a rifle bullet, or cannon ball. Two screens of tinfoil are set up, a short distance apart in front of the rifle, and these are made parts of two electric circuits through which currents pass to two electromagnets. The magnets hold back pens that are arranged to press upon a sheet of paper pasted on a drum which rotates by clockwork. So long as the screens are intact, the pens describe lines on the paper; but when a screen is broken, the corresponding pen instantly flies back and makes a nick in the line, thus:

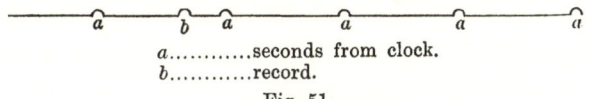

a............seconds from clock.
b............record.

Fig. 51.

If we know the distance between the screens, and the speed of the paper on the drum, and measure the lengths *aa*, *ab*, we can find the *average* speed of the bullet for the short distance between the screens; and by making this distance less and less we arrive at the conception of the *speed of the bullet at a definite point of its path.*

In the language of the Differential Calculus a very short distance is indicated by the symbol Δs, and a very short time by Δt. The speed at a given instant will then be the value of $\dfrac{\Delta s}{\Delta t}$ in the limiting case when both Δs and Δt are vanishingly small, and t includes the given instant. This limiting value is written $\dfrac{ds}{dt}$, and the notion of such a limit, as measuring the rate of change of some quantity s (in our case a distance) as compared with some other quantity t (in our case the time), is the fundamental idea out of which the Differential Calculus grew. We mention it here for the sake of its importance, though we shall not use it in our further calculations.

72. In the study of the falling stone there is another notion that we cannot do without. Such questions arise as these: Do all bodies in falling get up speed at the same rate? Does the same stone increase its speed by the same amount in each second of its fall? We are thus forced to the idea of the *rate at which the speed changes.* The name Acceleration is given to this idea, and it is convenient to apply it in cases where the speed is diminishing, as well as when it is increasing, considering the acceleration positive when the speed increases, and negative when it diminishes.

4. **Acceleration.**

Acceleration clearly bears to velocity the same relation that velocity bears to distance travelled. It may be measured on the same principles. If it be found that in t seconds a body has acquired a new velocity of v feet per second, the acceleration a will be given by

$$a = \frac{v}{t}.$$

As before, we see that v and t may be as small as we choose, provided we have the means of measuring them, without affecting the principle; and that a body may have an acceleration varying from moment to moment exactly as in the case of velocities.

73. Recurring to the experiments on the inclined plane, Galileo perceived that a falling body had a constant acceleration, the same at all parts of its fall, since in each second the speed increases by equal increments. Since we know that the speed acquired in one second is 32·2 feet per second, we can now calculate the speed acquired in any number of seconds from the formula

$$v = 32\text{·}2 \times t,$$

and the distance fallen through, from the formula

$$s = \frac{32\text{·}2}{2} \times t^2 = 16\text{·}1 \times t^2.$$

We may now sum up Galileo's investigations of the motion of a falling body in the statement that its weight produces a constant acceleration downwards, measured by the speed gained in each second (or lost, if the body is rising), viz. 32·2 feet per second. This acceleration is usually denoted by the letter g, and for rough calculations (unless otherwise stated) it may be taken as

$$g = 32\text{·}2.$$

74. One other point due to Galileo must be mentioned. Since the stone acquires during every second of its fall an extra speed of 32·2 feet per second, the action of the weight in producing speed is clearly independent of any speed the body may previously possess. For it gains as much speed during the third second, for example, when it starts the second with speed of 64·4, as during the first second, when it starts from rest.

5. The Path of a Projectile.

Galileo at once extends this to the case where a body has, to begin with, a velocity in *some other direction than the vertical.*

Let a stone be thrown horizontally from A with a speed v. Its motion will consist of two parts.

(1) As a heavy body, it will in t seconds fall through a vertical height $gt^2/2$, and if this were all, it would arrive at N, where

$$AN = gt^2/2.$$

(2)   But meanwhile the horizontal velocity will carry it to the right through a distance $AT = vt$, since nothing is hindering or

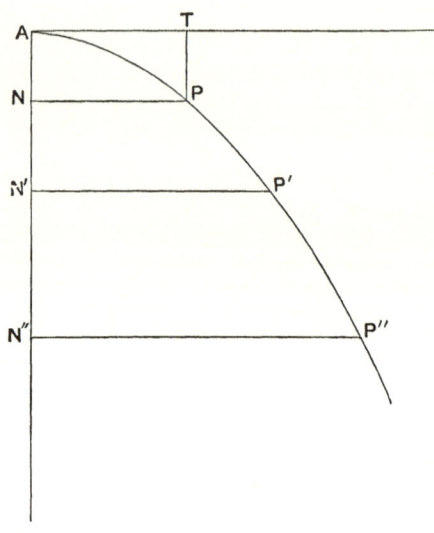

Fig. 52.

helping its horizontal motion, and therefore by the First Law of Motion (§ 70) its horizontal speed will remain unchanged.

Galileo sees that it will thus arrive at P, since both motions will take place simultaneously. Plotting a number of points such as P, corresponding to different times, and noting that in every case

$$PN^2 = v^2 t^2 = \frac{2v^2}{g} \times \frac{gt^2}{2} = \frac{2v^2}{g} \times AN,$$

he finds that the stone will describe a parabola.

We shall deal with the theory of projectiles more fully later on, and with another discovery due to Galileo, which can only be mentioned here, viz. the constancy of the time of oscillation of a pendulum, a discovery which Galileo was the first to apply to timing the pulse in disease.

EXAMPLES.

$(g = 32.)$

1. From observations of the eclipses of Jupiter's moons, which take place too early when the earth is between the sun and Jupiter, and too late when the earth is on the opposite side of the sun to Jupiter, Roemer concluded that light requires 16 minutes 36 seconds to cross the earth's orbit. Taking this diameter to be 195,600,000 miles, find the velocity of light in miles per second.

2. The velocity of sound in air at the ordinary temperature is 1120 feet per second. A thunder clap is observed to follow the flash of lightning at an interval of 8 seconds. How far off is the storm if the time required by the light to reach the observer is taken as zero?

3. Express a velocity of 60 miles an hour in feet per second.

4. A train 300 feet long passes a telegraph post in 12 seconds, and, gaining speed steadily, passes another post a quarter of a mile further on in 10 seconds. Find

(1)   the average speed at each post in miles per hour;

(2)   the average speed between posts;

(3)   the time taken to travel from one to the other;

and compare the acceleration of the train with that of gravity.

5. How long does it take a body to fall down a vertical precipice of 2000 feet?

6. A stone is dropped from a cliff into the sea, and is seen to reach the water in $4\frac{1}{2}$ seconds. What is the height of the cliff?

7. A stone dropped down a well is heard to strike the water after $4\frac{1}{4}$ seconds. The temperature being just above freezing point, the velocity of sound is 1024 feet per second. What is the depth of the well?

CHAPTER X.

75. GALILEO in the course of his investigations of the motion
of falling bodies, had been led to the following conclusions :—

(1) The weight of a body, if free to act, produces in it
a constant downward acceleration, such that it gains speed at
the rate of $g = 32 \cdot 2$ feet per second in every second of its fall.

(2) The distance fallen through in a time t seconds is $gt^2/2$.

(3) If the weight is counteracted, as when the body arrives
upon a smooth horizontal plane, its speed is no longer altered, but
it moves forward uniformly in a straight line.

(4) If the body is projected with any speed in any other
direction, the action of the weight is in no way affected by this
new speed, but the motion of the body is compounded of the
motion of a falling body and the uniform speed in a given direction
initially imparted to it.

76. The next case to be investigated was that of a body
moving round in a circle with uniform speed, as when a stone is
placed on a smooth horizontal table, so as to neutralize the effect
of its weight, and then whirled round at the end of a string. It
will be found that, when a fair speed has been attained, the hand
which holds the string is practically motionless at the centre
of the circle, but is conscious of a steady outward pull on the
string.

This problem was investigated by C. Huyghens (1629—1695)
some of whose many brilliant services to Mechanics must be more

D. CONSTANTINVS HVGENS EQVES
TOPARCHA SVYLE COM

Ant: van dyck pinxit   cum privilegio.

Paul Pontius fecit

fully noticed later on. Here we will give his solution of this particular problem in more modern form.

Once Galileo's ideas were abroad, it was natural to ask : Why does not the stone move onwards in a straight line with unchanging speed ? The answer is obvious. It is subject to a pull, applied by means of the string, of the same nature as the weight of the stone, since the weight might be supported by such a pull, and so balanced. Just as the weight of a falling body gives it a constant acceleration downwards, this pull must give the stone an acceleration along the string at each point of its path, by which it constantly acquires velocity towards the centre, and is deflected from the tangent at that point.

Before finding what this acceleration is, let us observe that a string cannot be stretched by means of a pull applied to one end only. The hand applies a pull inwards at the centre, and we are conscious of an apparent outward pull exerted by the moving stone. This is the so-called " centrifugal force." In studying the motion of the stone it must be remembered that the stone is subject to the pull *inwards*.

77. To find the acceleration a of a stone describing a circle of radius r with uniform velocity v.

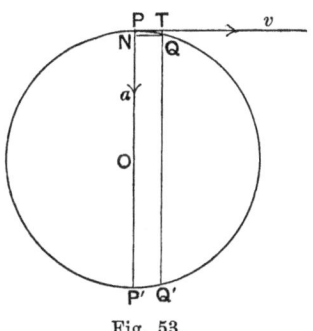

Fig. 53.

(1) *Direction.* The acceleration must be directed always towards the centre of the circle. For since there is no change in the speed with which the stone moves along the circle, any new velocity acquired must be at right angles to the circle at that point, *i.e.* along the radius.

(2) *Magnitude.* Suppose the string cut when the stone is at P. Then the stone would move along the tangent, with unchanged speed, and in a very short time t seconds, would travel a distance

$$PT = vt.$$

Again, suppose the stone to be at rest at P, and then acted on by the pull of the string for the same very short time t. By

Galileo's law of falling bodies, the acceleration produced, a, would cause it to fall through a distance

$$PN = at^2/2 = TQ.$$

If the time t be very short indeed, say one billionth of a second, the direction of the string, and therefore of the acceleration, will be parallel to PO throughout.

Now the actual motion of the stone is compounded of both these motions taking place simultaneously; and it arrives at Q.

Since Q is a point on the circle, the distances PN, PT are not independent, but are connected by definite geometrical relations. For instance, if TQ be produced to cut the circle in Q', we know that

$$TQ \cdot TQ' = TP^2.$$

Substituting the values of PT, TQ, we have

$$at^2/2 \times TQ' = v^2t^2.$$

When the time t is made very short, TQ' approaches more and more nearly to POP', and since the *form* of the law connecting TP, TQ remains the same, however small t is made, we see that when we arrive at the actual case of the stone changing its direction from point to point,

$$TQ' = PP' = 2r$$

and

$$a = \frac{2v^2t^2}{2r \cdot t^2} = \frac{v^2}{r}.$$

This may be put in another form which is often convenient, since we frequently know the radius of the circle, and the periodic time, that is the time in which the stone makes one complete revolution, so as to arrive at the point from which it started.

Let this time of one revolution be T seconds.

Then

$$v = \frac{2\pi r}{T},$$

and

$$a = \frac{v^2}{r} = \frac{4\pi^2}{T^2} \cdot r.$$

Since the radius to the stone describes in T seconds an angle whose circular measure is 2π, its *angular velocity*, measured in radians per second, is $\frac{2\pi}{T}$. Let this be denoted by ω. Then the acceleration may be expressed in still a third way,

$$a = \omega^2 r.$$

EXAMPLES.

1. Compare the "centrifugal force" on a body at the equator with gravity, taking the earth's radius as 3963 miles, and the time of revolution to be 86164 seconds.

2. A flywheel 10 feet in diameter makes 40 revolutions a minute. Compare the "centrifugal force" at the rim with gravity.

3. A train runs round a curve of radius 600 feet at 60 miles an hour. What acceleration must it receive inwards?

4. A stone is whirled round in a vertical plane at the end of a string 2 ft. 6 ins. long. Shew that in order that it may describe perfect circles, its velocity at the lowest point must not be less than 20 feet per second.

5. In a "centrifugal railway" the cars, after descending a steep incline, run round the inside of a vertical circle 20 feet in diameter, making a complete turn over. Shew that if there were no friction, they must start from a point not less than 5 feet above the top of the circle.

CHAPTER XI.

FINAL STATEMENT OF THE PRINCIPLES OF DYNAMICS.
EXTENSION TO THE MOTIONS OF THE HEAVENLY BODIES.
LAW OF UNIVERSAL GRAVITATION. NEWTON.

"Qui genus humanum ingenio superavit."

78. HUYGHENS' theorems concerning circular motion were
published in 1673 in his *Horologium Oscillatorium*, which contained many other discoveries of capital importance; both
geometrical, such as the properties of cycloids, evolutes, and the
theory of the circle of curvature; and practical, as the invention
and construction of the pendulum clock, the escapement, and the
method of determining the acceleration of gravity by means of
pendulum observations.

Galileo's ideas were evidently becoming familiar. It was
recognized that a motion would continue unchanged unless there
were some circumstance to interfere with it, or, as we should say,
unless some force acted on the body to alter its motion; that if
such a force acted, it would produce a change of speed, or an
acceleration, from which the motion could be calculated; and that
motion in a curve would result from the combination of a deflecting
force with a motion already existing, as in the case of projectiles,
and motion in a circle.

79. The problem that stood next in order for solution was the
most imposing and difficult that has ever been achieved by the
human mind. It was—To explain the movements of the heavenly
bodies, other than the fixed stars, *i.e.* of the moon, the planets,
their satellites, and the comets.

ISAACVS NEWTONVS

The first steps had already been taken. Copernicus had shewn that the complicated movements of the planets, which appeared to advance, stand still, recede, and then advance again with baffling irregularity, when viewed by a spectator on the earth, could be reduced to comparative order and simplicity, if the sun were regarded as the fixed point about which they took place. The planets, the earth being now one of them, apparently described circular orbits about the sun.

80. Closer observation shewed that this was not exactly the case. Next, Kepler with incredible patience had deduced the true laws of their motion from a life-long study of Tycho Brahe's observations. He found that :—

(1) The planets describe ellipses about the sun in one focus.

(2) The areas swept out by the radius vector in any orbit are proportional to the times.

And he had found, near the end of his life, a third law connecting the different orbits.

(3) The squares of the periodic times are proportional to the cubes of the semi-axes major (or of the mean distances).

81. Attempts had even been made to explain the planetary movements on mechanical principles, *i.e.* to trace a relation between them and such cases of motion as were already familiar. Thus Kepler himself suggested that the planets were carried round by spokes or radii attached to the sun ; and Des Cartes invented a theory of Vortices according to which each planet was maintained in motion by a whirl or eddy in a fluid which filled all space. But these guesses arose from a false idea—that motion required something to keep it up ; and this was contrary to Galileo's First Law of Motion. What was needed was, not a force to sustain their motion, but a deflecting force, that might cause them to move perpetually out of the straight line along the curve of their orbits. And since the ellipses they described differed but little from circles, and their speed in their courses varied very slightly, it looked as if any disturbing force must act almost at right angles to their direction of motion, and therefore towards the central body.

82. Seven years before Huyghens published his *Horologium Oscillatorium*, these ideas had been clearly grasped by a young graduate of Trinity College, Cambridge, who was destined to evolve from them the complete solution of the great problem, and to bring every known motion in the universe beneath the sway of a single law. As an incident in the course of his work, he completed the fundamental principles of Dynamics, and stated the Laws of Motion in a form which with the aid of proper mathematical analysis, suffices for the solution of all other mechanical problems.

Isaac Newton, by universal consent the greatest name in the roll-call of Science, was born on Christmas day 1642 at the Manor House of Woolsthorpe, a hamlet about six miles from Grantham in Lincolnshire. His father was a yeoman farmer. His mother, Hannah Ayscough, already widowed when Newton was born, is spoken of as "the widow Newton, an extraordinary good woman." One of her brothers held a neighbouring living, and was a graduate of Trinity College, Cambridge. Small hint in such ancestry of the genius which has impressed contemporaries and posterity alike as almost superhuman !

Newton was educated at Grantham Grammar School till the age of fifteen, and shewing no aptitude for farming, was on the advice of his uncle sent back to school and in 1661 to Cambridge. He graduated in 1664 ; was made Fellow of his College in 1667 ; and Lucasian Professor of Mathematics 1669, succeeding Barrow who had noted his unparalleled genius. Meanwhile as an undergraduate he had discovered the Binomial Theorem in Algebra, and had begun the invention of his method of Fluxions, now known as the Differential Calculus. 1665 was the year of the Great Plague. The whole College was sent down, and Newton returned to Woolsthorpe, there for a quiet year to ponder the new ideas he had gathered at the University.

83. It would be presumptuous to speculate on the workings of a mind like Newton's. Fortunately he has himself described the course of his thoughts, and there are other accounts by Wharton and by Pemberton based on his lectures and conversations in after years. Thinking over Kepler's Laws in the light of

Galileo's dynamical principles, he was led to see that the planets could be made to describe circular orbits with uniform speed, if they were acted on by a force emanating from the sun, whose intensity was inversely proportional to the square of the distance.

For since (§ 77) a body describing a circle of radius r in a time T has an acceleration to the centre

$$4\,\frac{\pi^2}{T^2}.r,$$

and since by Kepler's third law T^2 for the different planets is proportional to r^3, the accelerations must be inversely proportional to r^2. If the attracting force were looked on as an emanation from the sun, one might almost expect such a law of diminution of its intensity. For the areas of the spheres which have to be affected at greater and greater distances increase as the squares of their radii, and hence the intensity at any particular spot on any sphere would be inversely as the square of its radius.

84. But would the same law hold for the actual case, *i.e.* for elliptic orbits about the sun in one focus? In the first place Kepler's second law shewed that the force must still act towards the sun (§§ 225–6). To prove that the law of the inverse square was the correct law, and the only correct law, for an elliptic orbit described about a centre of force in one focus (not about the centre of the ellipse, for that requires a different law, § 229) was a mathematical feat that required the genius of a Newton. It was, in fact, the intellectual part of his achievement, and may be easily appreciated even now by any one who, with a fair training in Conic Sections, will try to work out § 232 for himself.

85. It is not so easy to realize, because the fact has become so familiar, the audacity of imagination by which Newton discerned, first in the case of the moon, that the force which regulates the courses of the heavenly bodies is no more than ordinary gravity, which pulls a stone to the earth. According to the well known anecdote the great flight of fancy was taken when Newton, convinced of the need of such a central force, directed towards the earth, to account for the moon's motion, but unwilling to adopt the theory till he could lay his finger on the force, observed the fall of an apple in the orchard at Woolsthorpe*. "The earth pulls the apple, though not connected with it. Why should it not pull the moon ?"

When you think of it, this is not so daring after all. For the earth pulls stones at all heights accessible to us. Why should it cease to do so even at the height of the moon ? We should even expect this according to the principle of continuity.

86.   But is it so ?   Newton made the simple calculation, using the data at his disposal.

The moon's distance is sixty radii of the earth.   Gravity at

the moon should be $\dfrac{1}{60^2}$ as

powerful as at the surface of the earth, (*i.e.* one radius from the centre).   At the surface of the earth it pulls a stone through $16t^2$ feet in t seconds, by Galileo's formula; therefore through 16×60^2 feet in a minute.   If it is gravity which holds the moon in its orbit, the moon should fall

through $\dfrac{1}{60^2}$ of 16×60^2, *i.e.* 16

feet per minute towards the earth.

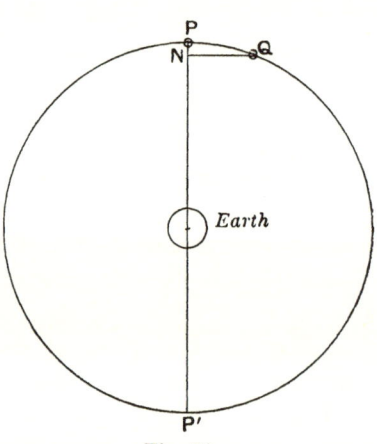

Fig. 54.

Let PQ be the distance travelled by the moon in one minute. Then the distance it falls towards the earth is

$$PN = \frac{QN^2}{NP'} = \frac{PQ^2}{PP'} \text{ approximately,}$$

$$= \frac{PQ^2}{120R},$$

where R is the earth's radius.

* We are permitted to believe the story, for it is explicitly stated to be the fact by Conduitt, his assistant at the Mint, and husband of his favourite niece ; by Voltaire, who had it from Mrs Conduitt ; and by R. Greene, on the authority of

But $$PQ = \frac{2\pi \times 60 \times R}{39343}$$

since the moon's period is 39343 minutes.

For the value of R Newton had only the nautical estimate that every degree was 60 miles, so that a circumference, or $2\pi R = 60 \times 360$ miles.

Thus

$$PN = \frac{2\pi \times 2\pi R \times 60^2}{120 \times 39343^2} = \frac{2 \times 3\cdot14 \times 60 \times 360 \times 60^2 \times 5280}{120 \times 39343^2} \text{ feet}$$

$$= 13\cdot88 \text{ feet.}$$

This is too small. If gravity were the force acting, the moon should fall through 16 feet per minute.

87. Newton's behaviour in face of this disappointing result is as marvellous an instance of scientific reserve, as his daring guess was of scientific imagination. At the age of twenty-three the secret of the universe is almost within his grasp. Nay, it is certain that Kepler's laws can be explained by a central force inversely as the square of the distance. But "hypotheses non fingo." Rather than base his theory on anything but a "vera causa," a force known to exist on other grounds, he lays aside the whole subject in silence.

Six years later, in 1672, Picard of Paris communicated to the Royal Society a new and careful determination of the size of a degree, making it 69½ miles instead of 60. When this came to Newton's knowledge, he took out his old papers, and made the calculation with the new numbers, and it is said that as it became clear that the result would accord with theory, his excitement was so great that he could not see the paper to finish his work*.

For two years he now threw himself into the labour of working out the detailed application of his discovery with such ardour and concentration of mind that he often forgot to take food†.

Martin Folkes, who was associated with Newton as Vice-President of the Royal Society when Newton was President. The tree from which, according to tradition, the apple fell, was blown down in 1820, and some of its wood has been preserved.

    * There seems to be no authority for this particular story, first given by Robison in 1804.

    † There are many anecdotes illustrating Newton's absentmindedness during these years.

88. A great deal had to be done. First a new method had to be devised, for treating things which were almost insensibly but yet continually varying, such as the direction of motion in the curved path of a planet, and its gradually changing speed; and the varying force which acted on it as it moved closer to the sun or farther away from it. From the doctrine of limiting ratios of vanishingly small quantities employed by Newton arose the Method of Fluxions and afterwards the Differential Calculus.

89. Next, the principles of Mechanics had to be collected and completed, and put in a shape convenient for the calculation of orbits described under any laws of force, and particularly that found in nature, the inverse square of the distance, about which a number of special propositions had to be proved. Then these had to be verified for the planets and their satellites, and the comets also brought under the law.

But if the earth, the sun, and the planets attract other bodies according to this law, why not also much smaller bodies such as meteorites? Again, must we not suppose that every part of the earth shares in producing the attraction? Nay, must not every part attract every other part? Or, rather, must not every particle in the universe attract every other particle according to the law of the inverse square of the distance?

90. Before it was possible to accept this idea, Newton had to shew how a sphere, such as the earth, composed of particles attracting according to the inverse square of the distance, would act on an external body. He worked out a number of propositions shewing how a sphere would act on a particle inside it, on its surface, and at any distance outside. For instance, in the last case the attraction is the same as if the whole substance were collected at the centre of the sphere.

91. Another application was based on Kepler's third law. Knowing the distance of any planet from the sun and the length of its year, and also the length of the lunar month and the distance of the moon from the earth, Newton was able to compare the mass, or "quantity of matter" in the sun with that of the earth.

92. But perhaps his most amazing achievement was his treatment of the perturbations of the moon's orbit. The moon is attracted by the sun as well as by the earth, and hence her motion varies in a number of ways from a true elliptic motion about the earth. The most important of these "inequalities" or irregularities were worked out by geometrical methods which no one has been able to advance beyond the point where Newton left them.

These are :

(*a*)   The evection, a periodical change in the eccentricity of the ellipse, discovered by Hipparchus and Ptolemy.

(*b*)   The variation, by which new and full moon occur a little too early, and the quadratures a little too late ; and

(*c*)   The annual equation, a variation in the other perturbations depending on the varying position of the earth in her orbit.

These two were discovered by Tycho Brahe.

(*d*)   The regression of the nodes ; and

(*e*)   The variation of the inclination of the moon's orbit.

These were at the time being observed by Flamsteed at Greenwich.

(*f*)   The progression of the apses, whereby the moon's orbit turns round in its own plane through $3°$ in a year.

Two other inequalities:

(*g*)   The inequality of the apogee,

(*h*)   The inequality of the nodes,

were predicted by Newton, never having been noticed by the observers before.

Newton's calculation of (*f*) gave only $1\frac{1}{2}°$, one half the observed amount. D'Alembert, Clairaut, and others of the great analytical mathematicians attempted to account for this curious discrepancy, but arrived at the same result; till at last Clairaut found that a number of terms had been omitted in the series as unimportant, which turned out to be not negligible, and when these were included, the result was correct. It was not till Professor Adams, one of the discoverers of Neptune, was editing the Newton papers in the possession of the Earl of Portsmouth, that MSS. were discovered shewing that Newton had himself reworked the calculations, and found out the cause of error, but had not published the correction !

93. Newton made several other remarkable applications of his theory.

From the time of revolution of the earth, considered as a fluid mass, he calculated its oblateness, and conversely, from the observed shape of Jupiter he estimated the length of Jupiter's day.

Then from the shape of the earth so deduced, combining its attraction with the effect of the " centrifugal force " of its rotation, he compared the force of gravity at the poles and the equator.

Again, from the attractions of the sun and moon on the earth's equatorial protuberance, he explained and calculated the precession of the equinoxes.

Finally, he worked out the theory of the Tides from the unequal attraction of the moon upon the solid nucleus of the earth, and on the nearer and farther parts of the ocean, taking account of special conformations of land and water in different parts of the earth; explained the spring and neap tides as the resultant of the tides due to the moon and the sun; reckoned the height of the solar tide from his known mass; and then from the observations of the spring and neap tides deduced a first estimate of the mass of the moon.

94. "In 1683, among the leading lights of the Royal Society, the same sort of notions about gravity and the solar system began independently to be bruited. The theory of gravitation seemed to be in the air, and Wren, Hooke, and Halley had many a talk about it."

"Hooke shewed an experiment with a pendulum, which he likened to a planet going round the sun. The analogy is more superficial than real. It does not obey Kepler's laws; still it was a striking experiment. They had guessed at a law of inverse squares, and their difficulty was to prove what curve a body subject to it would describe. They knew it ought to be an ellipse if it was to serve to explain the planetary motion, and Hooke said

he could prove that an ellipse it was; but he was nothing of a
mathematician, and the others scarcely believed him.   Undoubt-
edly he had shrewd inklings of the truth, though his guesses were
based on little else than a most sagacious intuition.   He surmised
also that gravity was the force concerned, and asserted that the
path of an ordinary projectile was an ellipse, like the path of a
planet—which is quite right*."

In January 1684 Wren offered a prize—a book worth forty
shillings—to Hooke and Halley if either of them could produce a
proof that a body under the law of inverse squares would describe
an ellipse.   But as nothing was forthcoming, in the following
August Halley made a journey to Cambridge and put the question
to Newton, who then for the first time told him that he had worked
it all out a dozen years before!   Halley communicated his discovery
to the Royal Society, and at their request Newton revised and
completed his papers and allowed them to be published.   The
Principia appeared in 1687, and in connection with this
momentous event it should not be forgotten that not only was its
publication brought about by Halley, but that he saw the work
through the press and defrayed the cost at his own risk.

95.   It is obvious that the conception of universal gravitation
introduces a wonderful simplicity and order into our views of the
varied and complicated motions of the heavenly bodies.   It sets us
at a new point of view from which the intricate movements which
had puzzled the ages are seen to fall into their places as parts of a
general scheme whose secret can be expressed in a single, simple,
universally applicable statement.

But Newton is very careful to warn us that he has not
discovered the causes of these movements; not even framed a
hypothesis to account for the attraction of gravitation.   He has
found a formula which describes with extreme simplicity and
universality *how the motions go on*.   So that by means of it we can
comprehend them in all their bewildering complexity, or communicate our knowledge to others, or calculate how they will go
on happening, and use our predictions to shape our conduct
wisely.   Not even the fall of a stone is explained, in the sense that

* Lodge, *Pioneers of Science.*

a cause is found for it. Nothing is explained in this sense, but
by the law of gravitation the most complicated motions known to
us in the universe are brought into relation with the simple case
of the stone, and what was remote and strange is reduced to what
is familiar enough.

96. We have dwelt so long on the discovery of the law of
gravitation partly because of its intrinsic interest and importance,
and partly for the sake of the light it throws on the method by
which science advances. But we must now turn to—what more
immediately concerns us—several very important steps in the
development of Mechanical theory to which Newton was led in
the course of solving his great problem.

97. Newton realized more clearly than had ever been done
before that motion could be altered not only by
1. Generalization
of the idea of
Force.
means of pushes and pulls, in which we are
conscious of the muscular effort by which we effect
the change, but by other circumstances, such as the supposed
attraction of the earth, and the known attractions of electrified and
magnetized bodies, which Dr Gilbert of Colchester had recently
written about so admirably.

Now when we make a muscular effort, we say that we exert a
Force. Newton generalized this idea so as to make it include all
the other cases, and gave it the definition still current.

Definition of Force. Force is any circumstance which changes
or tends to change a body's state of rest or of uniform motion in a
straight line.

2. The Parallel-
ogram of Forces.
98. This principle, already dimly grasped by
Galileo and Stevinus, was now stated explicitly.

99. In view of the law of gravitation it appeared that a body
3. The concept
of Mass.
might have very different weights according to its
position with regard to the earth, and indeed would
have no weight at all, if it were placed at the centre of the earth,
or at a very great distance from any attracting body. Nevertheless
the object remains unchanged even when its weight disappears.

We cannot remove objects to the centre of the earth or to immense distances. But let us neutralize the weight of some object. Thus a curling stone on smooth ice has its weight supported by the upward pressure of the ice. But a considerable effort is required to set it in motion, or to stop it when once started.

Hang two equal weights by a string over a smoothly running pulley. They will rest in any position. But an effort is required to get them into motion, and this effort will be greater the greater the size of the weights, and greater for leaden weights than for weights of equal size made of brass.

Again, a heavy fly-wheel on a smooth axle will rest in any position, but requires effort to start it or stop it.

From a consideration of such cases it is clear that there is something about a body, not its weight, which has a great effect in determining its behaviour when acted on by forces, and which, so far as we know, remains unchanged even when the weight is neutralized or altered. The term "Inertia" was introduced to indicate that bodies had no power to produce changes in their own motion, and offered an apparent resistance to changes of motion, which had to be overcome by an effort of some kind from outside.

There thus emerges a very important distinction between the weight of a body, which is variable and depends on its position with regard to some other attracting body, such as the earth; and something else, apparently an unchangeable attribute of the body, which determines how it will respond to the action of forces tending to change its motion.

Newton rather unfortunately called this "the quantity of matter" in the body. The modern term is "mass." We shall define this term more precisely later, but at present call the attention of the student to the distinction between mass and weight, which Newton was the first to realize.

100. So soon as it became necessary to calculate the move-
4. The Third Law ments of two bodies each of which attracted the
of Motion. other, the question arose "What is the relation
between the two mutual attractions?" Newton's answer was

that the two attractions would be equal; and he generalized the statement for forces of all kinds in his Third Law of Motion :—

" To every Action there is an equal and opposite Reaction."

This was perhaps the only absolutely new point in his statement of the laws of motion. These laws have proved sufficient for the solution of all problems in Dynamics. All that has happened since has been a mere matter of mathematical development of their consequences.

With the Laws of Motion the historical evolution of the fundamental principles employed in Mechanics may be considered completed.

BOOK II.

MATHEMATICAL STATEMENT OF THE PRINCIPLES.

INTRODUCTION.

101. WE have seen in Book I. how the fundamental principles of Mechanics were gradually won from experience by men of genius engaged in attempting to solve problems that either forced themselves on their attention by the practical importance of their consequences, as in the case of the Simple Machines, or attracted them by their own impressive grandeur and their bearing on questions of philosophy.

Some of these principles can be mutually deduced from each other. They are equally valid as results of experience, and would serve equally well as the starting point of the subject. The particular order in which they arose was largely a matter of historical accident.

We shall now give a more connected and precise statement of the subject, selecting that order which, after Newton, makes the Laws of Motion the starting point in experience of all the rest.

And as we are to study motion as produced by force in real bodies, let us begin by clearing up our ideas about motion by itself, apart from any consideration of what is moving or why it moves. This power of abstraction, or attending to one thing at a time, is of great value in science.

Let a small mirror be held in the sun so that a spot of light is reflected on to the walls, ceiling, and furniture of a room. Even a perfectly regular turning of the mirror will set the spot moving with baffling variations both of speed and direction. When it passes from wall to ceiling, there is an instantaneous change of direction such as never happens to a heavy body, no matter what the force applied to it. We could study the position, velocity, and change of velocity of such a spot, without any reference to the laws of motion according to which motion is found to arise and change in real bodies.

This part of the subject, which deals with pure motion in the abstract, is called Kinematics.

The part dealing with the motion of real bodies under the action of forces is called Kinetics.

The two branches together are often called Dynamics. The special cases where the forces concerned happen to be in equilibrium are usually classed together under the title Statics.

CHAPTER XII.

KINEMATICS.

102. CLERK MAXWELL says :

"The most important step in the progress of every science
is the measurement of quantities. Those whose
Measurement. curiosity is satisfied with observing what happens
have occasionally done service by directing the attention of others
to the phenomena they have seen; but it is to those who endeavour
to find out how much there is of anything that we owe all the
great advances in our knowledge."

In order to measure, or express the exact amount of any
quantity, two things are required : (1) a unit, or standard quantity
of the same nature as that to be measured ; and (2) a number to
indicate how many such units or parts of a unit are contained in
the given quantity. Thus a sum of money may be expressed as
20 shillings, or 4·86 dollars. But it would be impossible to convey
to a stranger any idea of a sum of money unless we had previously
come to an understanding as to the purchasing power of some
definite amount, such as a shilling or a dollar, and could both of
us count.

The number required to express any given amount will
evidently be greater the smaller the unit we employ, and *vice
versâ*. *The measure of a quantity is inversely proportional to
the unit in which it is expressed.*

103. All the civilised governments have united in establishing
an International Bureau of Weights and Measures
Length. in the Pavillon de Breteuil, in the Parc of St Cloud
at Sèvres, near Paris. Here are kept the standards of length and
mass.

The unit of length is the International Metre, which is defined as the distance, at the melting point of ice, between the centres of two lines engraved upon the polished surface of a platiniridium bar, of a nearly X-shaped section, called the International Prototype Metre. The international metre is authoritatively declared to be identical with the former French metre, or mètre des archives. This was intended to be one-ten-millionth part of a quadrant of a terrestrial meridian. But as the value of a quadrant came to be more accurately determined, and moreover is changing, the actual bar constructed has been made the standard, and succeeding determinations of the quadrant are now expressed in terms of it.

The accepted unit of length in all scientific works except those of British engineers, is the centimetre, or one-hundredth part of the standard metre.

The British unit of length is the Imperial Yard which is the distance at 62° F. between the centres of two lines engraved on gold plugs inserted in a bronze bar usually kept walled up in the Houses of Parliament at Westminster.

The foot, or third part of the standard yard, is often employed as the unit in British works.

For measuring great distances multiples of these units are used, such as the kilometre and the mile; very small lengths are

A DECIMETRE DIVIDED INTO CENTIMETRES AND MILLIMETRES.

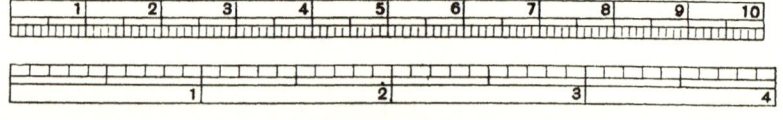

INCHES AND TENTHS.

Fig. 55.   1 metre = 39·37079 British inches.

often expressed in submultiples such as the micron, or one-thousandth part of a millimetre, *i.e.* the millionth part of a metre.

The relation between the two units is shewn in Fig. 55.

104. The universal unit of time is the mean solar day or its
one 86400th part, which is called a second.

Time.

This is the time during which the earth turns
on its axis through a certain small angle with reference to the
fixed stars.

Any time is measured by the number of seconds it contains.

It will be seen that in measuring the speed of any body by the
number of units of length it travels over in a unit of time, we are
really comparing its motion with another motion, viz. that of the
earth on its axis, just as we compare a length with another length.
Should there be any change in the standard length or in the rate
of the earth's rotation on its axis, our measures would lose their
meaning. Clerk Maxwell suggests that those authors who think
their works likely to outlast the present condition of the earth,
would do well to express their lengths in terms of the wave-length
of some particular ray in the spectrum, and times in terms of the
periodic time of vibration of such a ray, quantities which we have
every reason to believe will remain constant so long as the physical
universe retains its identity.

At the request of the French Government, Professor Michelson
has determined the value of the standard metre in wave-lengths
of the red, green, and blue rays of cadmium, and finds that

$$1 \text{ metre} = 1{,}553{,}163 \cdot 5 \text{ wave-lengths} \ldots \ldots \text{ red} \quad \text{ray}$$
$$= 1{,}966{,}249 \cdot 7 \qquad \text{,,} \qquad \ldots \ldots \text{ green } \text{,,}$$
$$= 2{,}083{,}372 \cdot 1 \qquad \text{,,} \qquad \ldots \ldots \text{ blue } \quad \text{,,}$$

at 15° C. and 760 mm. pressure.

105. The position of a point in a straight line is fixed when
we know its distance from some fixed point of

Position.

reference, or origin, in the straight line, and the
direction in which this distance is to be measured.

It is convenient to prefix the positive sign to all distances
measured in one direction, *e.g.* towards the right if the line is
horizontal, and the negative sign to all distances measured in the
opposite direction (towards the left if the line is horizontal), from
their respective starting points. A distance will thus be positive,
even though it lies entirely to the left of the origin, provided it
is measured towards the right.

With this convention, the algebraic sum of all the distances (with their proper signs) travelled by a point starting from the origin will always give us its position, and the sign of the sum will tell us whether it is to the right or left of the origin.

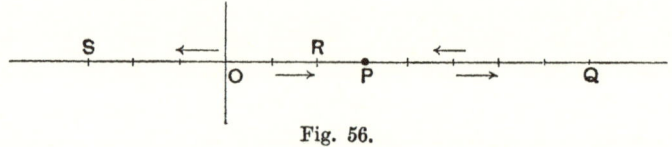

Fig. 56.

Thus if a man starts from a town O, and walks 3 miles due east, his position at P is indicated by $+3$. If he now walk 5 miles more to the east, he will arrive at Q, where $OQ = +3 + 5 = +8$.

Let him now walk 6 miles westward. He will then be found at R where

$$OR = +3 + 5 - 6 = +2.$$

Finally let him continue westward for 5 miles. He will be at S, where

$$OS = +3 + 5 - 6 - 5 = -3.$$

106. The position of a point in a plane may be fixed by Cartesian Co-ordinates, so called after their inventor Des Cartes, or by Polar Co-ordinates.

(1) Cartesian Co-ordinates.

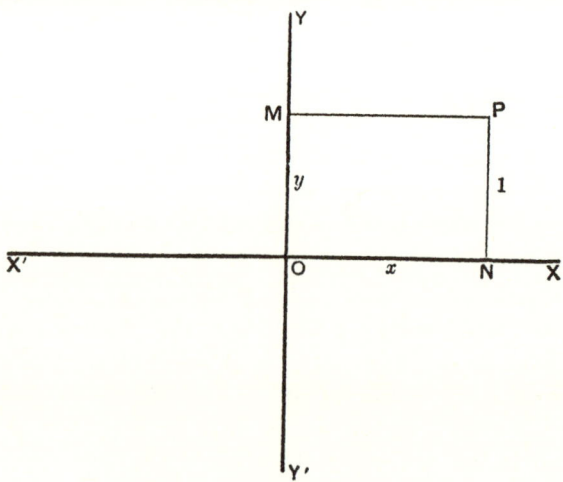

Fig. 57. Cartesian Co-ordinates.

Two lines of reference, XOX', YOY', are chosen, usually at right angles. These are called *the axes*. Through any point P parallels to the axes are drawn, cutting off lengths ON, OM. These are *the co-ordinates* of the point P. Lengths along YOY' are counted positive if measured upwards, and negative if measured downwards, from their starting points, whether the lengths themselves be above or below O. When ON, OM are known, the position of P is fixed by lines drawn through N and M parallel to the axes.

(2)   Polar Co-ordinates.

The position of P may also be fixed by the distance OP and the angle $POX = \theta$ through which OP must revolve from OX to reach P.

Angles turned through in the opposite direction to that of the hands of a clock (counter-clockwise) are counted positive. Those turned through in the clockwise direction are negative.

Since the point P is considered as carried through the angle θ by the revolving radius OP, OP is called *the radius vector*. The radius vector and the angle θ are called *the polar co-ordinates* of P with reference to the pole O.

Two systems are employed for measuring angles in works on Trigonometry. The unit angle in one of them is the right angle, with its subdivisions into degrees, minutes, and seconds.

In the other, the system of Circular Measure, the unit angle is the radian, *i.e.* the angle subtended at the centre of any circle by an arc equal to its radius. This is an angle of about $57°\ 19'$.

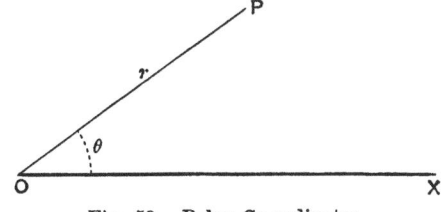

Fig. 58.   Polar Co-ordinates.

This system is often convenient in Mechanics, especially for dealing with rotations, for the length s of the arc of a circle of radius r described by P when the radius vector revolves through an angle whose circular measure is θ, is given by

$$s = r\theta.$$

107. *Definition.* A point is said to move when it changes its
Motion. position with reference to surrounding objects, or some particular object chosen for reference. The only motions known to us are thus relative motions. But this is no restriction, since in practice we are only interested in relative motions. We want to know for instance how to avoid a collision with another ship, or to strike a fort with a shell, and for all such purposes a knowledge of relative motion suffices.

For most purposes motions are referred to the surface of the earth, in spite of its own rapid and complicated motion. In astronomy the sun is chosen for reference so long as we confine ourselves to the solar system. The motion of the sun itself is referred to the general body of so-called fixed stars.

108. *Definition.* The *velocity* of a point is the rate at which
Velocity. it is changing its place.
The *unit velocity* is that of a point which passes over a unit length in a unit time. It is usually therefore a velocity of one foot or else one centimetre per second.

Any other velocity is measured by the number of units of velocity it contains. This is the same as the number of units of length passed over in a unit of time, if the velocity be uniform. This may be determined (§ 71) from the distance travelled in a very short interval of time, without waiting for a whole second.

There is no difficulty in extending this notion to the more usual case where the velocity is variable, *i.e.* is changing from moment to moment. Let the interval of time considered be chosen smaller and smaller, but always so as to include the particular instant at which the velocity is to be estimated. The numbers representing the distance travelled and the time occupied in traversing it may thus be made vanishingly small, but their ratio, which measures the velocity, has a finite value however small they are made. Common experience has familiarized the idea of speed *at* a particular moment, independently of the length of time for which it is maintained. Thus a train may pass a particular signal post at sixty miles an hour, and stop a few hundred yards beyond it. Everyone understands what is meant by the statement, and knows that if the speed had been maintained unaltered, the train would have covered sixty miles in the next hour.

109. Since a velocity is specified when we know its magnitude and its direction, it may be represented by a straight line. For the line may be drawn in the given direction, and of such a length as to represent the magnitude on any convenient scale. It is convenient to use the term "speed" to indicate the magnitude of a velocity irrespective of its direction, velocity implying that the direction also must be taken account of.

Geometrical representation of velocities.

110. *Definition.* The *acceleration* of a point is the rate at which it is changing its velocity.

Acceleration.

The *unit acceleration* is that of a point which gains one unit of velocity in one unit of time. It is usually therefore an acceleration of one foot per second in a second, or of one centimetre per second in a second.

Any other acceleration is measured by the number of units of acceleration it contains. This is the same as the number of units of velocity gained in a unit of time, if the acceleration be uniform. This may be determined from the velocity gained in a very short interval of time, without waiting for a whole second.

The notion may be extended to variable accelerations, exactly as in the case of velocities, by considering the velocities gained in a vanishingly short interval of time.

Acceleration thus stands to velocity as velocity stands to distance travelled, and, like velocity, may be represented by a properly drawn straight line.

111. We require formulae expressing the connection between the acceleration a, the velocity v at any time, the distance travelled s, and the time elapsed t.

The kinematic formulae.

(1) Uniform Velocity.

In this case there is but one formula, but care must be

taken on no account to use it in questions involving variable velocity.

If a point move for t seconds with uniform speed v, the distance travelled s is given by

$$s = vt,$$

with its equivalents,

$$v = \frac{s}{t}; \quad t = \frac{s}{v}.$$

112. (2) Uniform Acceleration.

Let a be the acceleration. Then if the point start from rest, it will acquire a units of velocity in every second, and at the end of t seconds it will have a velocity v, where

$$v = at \quad \dotfill (1).$$

To find the distance travelled.

The velocity at the middle moment (not the middle of the path), *i.e.* after $t/2$ seconds have elapsed, is $\frac{at}{2}$.

Compare the actual motion of the point with the motion of a point which starts at the same instant with the velocity $\frac{at}{2}$, and maintains its speed unchanged throughout. For every moment in the first half of the time when the first point is moving more slowly than the second, there is a corresponding moment in the second half when it is moving just as much faster. So that in the end the two points will cover the same ground.

Therefore

$$s = \text{(average velocity)} \times \text{(time of motion)}$$

$$= \qquad \frac{at}{2} \qquad \times \qquad t$$

$$= \tfrac{1}{2}at^2 \quad \dotfill (2).$$

Another form may be given to this result, which is important when we wish to connect the velocity acquired directly with the space passed over.

As above,

$$s = \text{(average velocity)} \times \text{(time of motion)}$$

$$= \qquad \frac{v}{2} \qquad \times \qquad \frac{v}{a} \qquad \text{(since } v = at\text{)}$$

$$= \frac{v^2}{2a}.$$

We write this $\qquad \dfrac{v^2}{2} = as$ ....(3).

113. If the point, instead of starting from rest, has an initial velocity u, these formulae admit of a simple modification. The effect of the acceleration in producing either new velocity, or extra distance travelled, or extra half-square of the velocity, has simply to be added to what is due to the initial velocity.

Thus, corresponding to

(Formulae for point initially at rest)		(Formulae, initial vel. u)
$v = at,$	we have	$v = u + at,$
$s = \frac{1}{2} at^2$		$s = ut + \frac{1}{2} at^2,$
$\dfrac{v^2}{2} = as$		$\dfrac{v^2}{2} = \dfrac{u^2}{2} + as.$

Otherwise thus:

From the definition of acceleration

$$a = \frac{v - u}{t}, \qquad \therefore \ v = u + at.$$

Again, the average velocity in this case is

$$\frac{1}{2}(u + \overline{u + at}) = u + \frac{at}{2}.$$

Now $\quad s = (\text{average velocity}) \times (\text{time of motion})$

$$= \quad \left(u + \frac{at}{2}\right) \quad \times \quad t$$

$$= ut + \frac{at^2}{2}.$$

Also $\quad s = (\text{average velocity}) \times (\text{time of motion})$

$$= \quad \frac{v + u}{2} \quad \times \quad \frac{v - u}{a}$$

$$= \frac{v^2 - u^2}{2a}.$$

$$\therefore \ \frac{v^2}{2} = \frac{u^2}{2} + as.$$

EXAMPLES.

1. Remembering that a velocity of 60 miles an hour is equivalent to 88 feet per second, write down in feet per second velocities of 15, 20, 36 miles an hour; and in miles an hour velocities of 8, 11, and 40 feet per second.

2. It takes light 3·315 years to come from the star a Centauri to the earth. If the velocity of light is 186,000 miles per second, and the radius of the earth's orbit is 92,370,000 miles, express the distance of a Centauri in radii of the earth's orbit.

3. Taking the earth's orbit to be circular, find the mean velocity of the earth in its orbit.

4. Find the velocity of a point at the equator due to the rotation of the earth on its axis, if the earth's radius is 3963 miles.

5. A bullet is fired through two screens 1 metre apart, and the interval required to pass from one to the other, as recorded on a chronograph, is ·0036 second. Express its velocity in centimetres and feet per second.

6. A steamer approaching a coast with vertical cliffs in a fog whistles, and the echo, as timed by a stop-watch, is heard after $8\frac{2}{5}$ seconds. One minute afterwards she whistles again and the echo is heard after $4\frac{2}{5}$ seconds. How far is she then off shore? How fast is she going? How soon will she strike if she goes on? (Sound travels a mile in 5 seconds.)

7. An enemy's guns are heard $4\frac{3}{5}$ seconds after the flash. Express the range in yards, assuming the velocity of sound to be 1120 feet per second.

8. A military band marches off at the rate of 9 steps in 5 seconds, covering 2 feet 6 inches every step. When they have made 122 steps after passing a spectator, they appear to him to be exactly out of step with the music. What was the velocity of sound that day?

9. A train moves from rest and after one minute has a velocity of 30 miles an hour. What is its acceleration?

10. A body starts with velocity 30 and after 8 seconds has velocity 90. What is its acceleration?

11. A body has acceleration 32 and starts with velocity 80. What is the velocity after 1, 4, and 10 seconds?

12. A stone dropped from a stationary balloon reaches the ground in 24 seconds. What was its velocity at the ground?

13. A stone is thrown vertically upwards with a velocity of 120 feet per second, the acceleration being 32 downwards. How soon will it be stationary? Find its velocity after 3, 4, and 7 seconds.

14. What is the acceleration of a train whose speed increases from 20 to 30 miles an hour in 100 yards?

15. A train running 40 miles an hour has the brakes put on and reduces its speed to 20 miles an hour in 220 yards. What is the acceleration? How much farther will it run before it stops?

16. (a) A bullet acquires a speed of 1600 feet per second while traversing a rifle barrel 4 feet long. Find the average acceleration.

(b) The muzzle velocity of a revolver bullet is 600 feet per second, and the barrel is 8 inches long. Find the average acceleration.

17. A point moves 12 feet in 1 second and 18 feet the next. How long has it been moving with uniform acceleration from rest? What is its acceleration? How far will it go in the next 10 seconds, and when will its velocity be 81 feet per second?

18. A body has initial velocity u and acceleration a. Find a formula for the space passed over in the nth second.

19. How long must a body travel with the acceleration of gravity before it acquires the velocity of light? How far would it move in the time?

20. A bullet is fired vertically upwards with a velocity of 1600 feet per second. After how many seconds will it return to the earth? What is the greatest height reached?

21. How far must a body fall to acquire a speed of 400 metres per second?

22. A velocity of 15 foot-second units is changed into 5 units while the body travels 50 feet. What is the acceleration? What would it have been if the velocity had been changed to −5 in the same distance? How much longer must the acceleration have acted in the second case, and how much farther will the body have travelled?

CHAPTER XIII.

KINETICS OF A PARTICLE MOVING IN A STRAIGHT LINE. THE LAWS OF MOTION.

114. In dealing with the movements of real bodies about us produced by our own muscular efforts we know from experience that the effect produced by a given effort will depend largely on the body to which it is applied.

So long as we are concerned with portions of the same substance, the size of the body determines the result, at all events approximately. Thus if a certain effort is required to project a stone with a certain speed, something like twice the effort will be needed to project in the same manner a stone of twice the size.

But when we compare the effects of the same effort on two bodies of different nature, such as cork and lead, something else besides mere volume has to be considered.

Newton gave to this "something" which determines the effect of an effort upon a body, the name "quantity of matter" in the body; rather unfortunately, as has been said, because the definition seems to raise the question "what is matter?" a question which has occupied philosophers from the earliest times without yet receiving a generally accepted answer. Newton himself immediately has to define how the "quantity of matter in a body" is to be estimated. He says that it is to be taken as the "product of the volume of the body and its density." But as the only way of determining the density is first to find the quantity of matter in a unit of volume, it is clear that we are landed in a logical circle.

Fortunately the physicist is not bound to enter on the thorny paths of philosophy, at least at this point. He may content himself with the answer of the Oxford undergraduate, who, when asked "What is mind?" replied "No matter"; "What is matter?" " Never mind."

And yet we cannot study scientifically the movements of real bodies without being able to measure this quantity which determines the motion that will be produced in them by a given effort. This implies that we shall be able to choose a unit quantity, and count the number of such units contained in a body, whether of cork, or lead, or any other substance.　How is this to be done?　The mere volume, we have seen, will not help us to deal with different substances.

For shortness, and to avoid the misleading associations of the word matter, let us call the quantity, unchangeable so far as we know, which determines the effect of a given effort in producing speed in a body, the *Mass* of the body.

And as Newton extended the ideas connected with our muscular efforts to all other cases where motion is changed by any means whatever, let us adopt his definition of Force.

Force is anything which changes or tends to change a body's state of rest or of uniform motion in a straight line.

It might be supposed that though we cannot compare masses by comparing their volumes, yet we might do so by comparing their weights.　And this, as Newton points out, happens to be true enough, at all events for comparisons made at the same place. But we must not assume that the weight of a body, which varies from place to place, and would be nothing at all at the centre of the earth, and only one-sixth as great at the surface of the moon, is a safe guide in measuring its mass, which no physical circumstances known to us will suffice to change in the slightest degree. We must find some other test of the equality of two masses.

If our object were to study the chemical properties of bodies, we might legitimately define equal quantities of two different substances as those which could neutralize the same amount of some standard reagent, such as sulphuric acid, if we could find one which acted on all substances; and the science could be logically built on such a definition.

Our actual purpose in Mechanics is to study the effect of forces in producing motion.　The proper test of the equality of two masses is therefore to observe whether the same force produces the same mechanical effect on them. This indeed is the test we instinctively apply in practice. If, of two equal casks lying on a wharf, one is known to be full and the other empty, and we wish to find out which is which, we give each of them a kick or push, and the one which resists us most is the full one.

Definition. Two masses are equal if the same force acting on each of them for the same time produces in each the same velocity.

To be sure that we are applying the same force we might apply it by means of a spiral spring, taking care to pull it so that its extension remains the same throughout. If we may assume that the physical properties of the spring remain unchanged, then the force will be constant. Though this is not a practical form of experiment, it is theoretically sufficient, and this is all that is necessary for our present purpose, which is to conceive a test by which other masses may be set off equal to the standard.

115. Two such units or standards are employed.

(1) The International Kilogramme, which is the mass of a certain cylinder of platiniridium kept at Sèvres, and intended to be identical with the former French kilogramme des Archives.

In science it is generally the Gramme, or thousandth part of the kilogramme, which is taken to be the unit. The gramme was intended to be the mass of a cubic centimetre of water at its temperature of maximum density 3·93° C.

The system of units which is based on the centimetre, gramme, and second, as units of length, mass, and time respectively, is called the c.g.s. system.

(2) The Pound, which is the mass of a certain platinum weight, called the British Imperial Pound.

According to Miller's determination

$$1 \text{ pound} = 0\cdot4535926525 \text{ kilogramme},$$

$$1 \text{ kilogramme} = 2\cdot204621249 \text{ pounds}.$$

The mass of any other body is expressed by the number of such units of mass (pounds or kilogrammes) it contains.

116. It is so important to keep clearly in mind the distinction between the mass and the weight of a body, and also the method of testing the equality of two masses, that we will consider an example in detail.

Suppose we wish to purchase a pound of sugar. Note that from Newton's point of view (mass = quantity of matter) we should be glad to have the mass as large as possible; whereas the weight, *i.e.* the pull with which it tends towards the earth, is purely an inconvenience when it comes to carrying the sugar home, and we might be glad if it could be done away with, provided that the mass were not thereby diminished.

To determine what is a pound of sugar, the shopman might set upon a long counter two little wheeled cars exactly alike in all respects. In one of these, according to our test, he should place a standard pound ; in the other a quantity of sugar. They should then be successively drawn along by means of a spring balance, care being taken that the reading of the balance always remained the same, and their progress timed. If the sugar were found to out-run the standard pound, more should be added ; if it fell behind, some taken away, until at last both cars were found to gain speed at the same rate. We should then have exactly one pound of sugar.

A simpler way would be first to test two spring balances by locking them together, and observing whether their readings were equal when they were pulled apart. Then, drawing the two cars along simultaneously, one by each balance, we could observe which of them required the greater force to gain approximately the same speed, and adjust the amount of sugar till the balances gave the same reading during the experiment.

Observe that the pound is not a force but a mass. The word is often employed to denote the force with which a pound tends downwards, *i.e.* the weight of a pound. When we specify a force by the number of pounds it would sustain, it is better always to use the correct but more cumbrous form, and speak of a force of so many pound-weights, except when there can be no possible danger of confusion.

Having defined Force and Mass and seen how masses may be measured, we are ready to study the laws of motion as stated by Newton.

117. Newton's Laws of Motion.

Law I. *Every body continues in its state of rest or of uniform motion in a straight line, except in so far as it is compelled to change that state by impressed force.*

This is merely Galileo's principle of Inertia, by which is meant that a body has no power in itself of altering its own state of motion, whatever that may be, but can only change it in response to some force applied from outside. As we have seen (§ 70), Galileo arrived at the law from the principle of continuity applied to the motion of a body down an inclined plane. It is distinctly contrary to the views generally held in his time, and by unobservant people to this day.

118. The full establishment of a principle of this kind generally consists of four stages :

I.   Observation ⎫
II. Experiment ⎬ . Induction of the Law from facts.

III. Deduction of Consequences of the Law.

IV. Verification, by comparison of the consequences deduced with further observations of facts.

The first of these stages is the most difficult, requires the greatest originality. To descry a new meaning in a fact whose very familiarity blinds our eyes to its significance, like Columbus with the drift-weeds on the western shore; to break away from inveterate prejudice to new points of view, like Copernicus; to catch in a flash of intuition the resemblance between remote facts, as when Bradley in his moving boat saw that the slant in the rain and the infinitesimal shift in the place of the fixed stars were akin, and so unravelled the aberration of light; to divine in the falling apple the secret of the heavens; this is the work of genius, of the poet's imagination, vivid in observation, fertile in surmise. Before a problem guesses come to most men, but in what surpassing measure to a Newton !

I.   Once the guess is made, it must be tested. Sometimes, as in Astronomy, we can do no more than wait till the event can be observed again, as with the phenomena connected with total eclipses of the sun, and transits of Venus.

II.   Or we may have the facts under our control, and be able to repeat them at pleasure, varying the circumstances.   Then we *experiment*, seeking to disentangle what is essential from what is indifferent, according to the canons laid down in works on logic.

These two stages, Observation and Experiment, constitute the Induction of the law from the facts.

III.   In deducing the consequences which should follow, if the law be true, the instrument most generally employed in Physics is Mathematics, which is only a systematic method of applying common sense with ease and accuracy.

IV.   Finally the results of calculation are carefully compared with fresh observations.   It not unfrequently happens that the theory leads to recondite consequences that would hardly have been stumbled on without its aid.   Thus Fresnel's Undulatory Theory of Light enabled Sir W. R. Hamilton to predict the conical refraction in crystals, afterwards observed by Thomas Young working from his directions; and Adams and Leverrier discovered Neptune from a consideration of the disturbances in the orbit of Uranus.   Such startling and dramatic verifications lead to the rapid adoption of a principle, but its final acceptance depends on patient comparison of calculation with observation resulting in universal agreement in detail.

119.   The First Law of Motion cannot be observed directly, because we cannot screen a body from the action of all forces and watch its behaviour.   Nevertheless the stage of observation was fulfilled when Galileo divined it from motion on an inclined plane.

For experiment, the more we do to remove retarding forces, the longer motion continues.   Push a table along a rough floor. It comes to rest (through friction) the moment we cease pushing. Place its legs on castors, and it will run a few inches after we let go.   Set the wheels on rails, as in a railway truck, and once started it will travel a considerable distance.   A block of ice thrown along a sheet of ice travels a very long way.   Two equal weights suspended by a fine thread over a very lightly running pulley (Atwood's Machine, Fig. 61 *a*) balance each other.   But if set in motion, the system travels with almost uniform velocity for a long time.

The inventions of the Perpetual Motion seekers are often good instances of approximation to the case of the First Law.

The stages of Deduction and Verification for all the Laws of Motion find a superb illustration in the *Nautical Almanac*. This volume of 600 pages, published four years in advance, contains on every page many hundreds of predictions of the places of the sun, the planets, the satellites of the planets, and of the moon among the fixed stars; and the dates, durations, place of commencement, path, and conclusion of eclipses, worked out to a degree of accuracy within the limits of error of the most sensitive modern instruments of precision. Every calculation is founded on the three Laws of Motion, applied to the averages of long series of corrected previous observations. Yet such is our confidence in their truth, that every ship captain unhesitatingly stakes his vessel on the results deduced from them; and it is safe to say that if an astronomer, provided with the finest instrument in the world, observed even a minute departure from its calculated place in one of the heavenly bodies, it would never occur to him to doubt the laws of motion, but he would search for some unusual source of error in his instrument, or suspect a new and undetected cause of disturbance, as did Adams and Leverrier in the case of Neptune.

120. The First Law states that unless some force acts on a body from without, its motion continues unchanged.

The Second Law tells us how the motion will be changed when a force acts on the body.

Law II. *Change of motion is proportional to the impressed force, and takes place in the direction of the force.*

By *motion* Newton does not mean *velocity only*, since the same force will produce very different changes of velocity in different masses. In measuring the *quantity of motion* we must therefore take account of the mass moved as well as of the speed produced.

We choose for *unit quantity of motion* the quantity of motion contained in one unit of mass moving with unit velocity. On the

British system this will be the quantity of motion possessed by one pound moving with a speed of one foot per second.

M pounds moving one foot per second will contain M times as much, and if the M pounds are moving V feet per second, there will be V times as much again. So that M pounds moving V feet per second contain a quantity of motion represented by MV such units.

It is time to have a single name for this recurrent phrase *quantity of motion.*

Definition. The quantity of motion in a body is called its *Momentum.* It is measured by the product of the mass of the body into its velocity. No special name has been given to the unit of momentum.

The word *proportional,* in Law II, is to be taken in its strict mathematical sense ; *i.e.* questions on the Second Law are to be worked out by Rule of Three, or Proportion.

In measuring the *impressed force* we must take account not only of the magnitude of the force, but also of the time during which it acts; since the longer a force acts the greater is the change of motion it produces.

The total effect of a force in producing change of motion is called its *Impulse* (*i.e.* total push).

We choose as *unit impulse* the effect of unit force acting for unit time, *i.e.* for one second.

P units of force acting for one second produce P units of impulse; and if they continue acting for t seconds, there will be Pt units.

121.   We can now state the Second Law as follows :

Momentum produced is proportional to the Impulse of the Force acting, and is in the direction of the force.

In algebraical symbols :

$$MV \propto Pt.$$

The sign of variation, $\propto$, is inconvenient in this equation, and may be got rid of by a proper choice of units. Now we have already chosen the units of mass, velocity, and time. But nothing has been said about the unit force. It is therefore open to us to choose a unit force, and we define it thus.

Definition. The unit force is that force which acting on unit mass for unit time produces in it the unit velocity.

Let us calculate, by rule of proportion according to the Second Law, the velocity that will be produced when P units of force act for t seconds on M units of mass.

By definition :

1 unit of force acting on 1 unit of mass for 1 second produces 1 unit of velocity.

P units of force acting on 1 unit of mass for 1 second produce P units of velocity.

P units of force acting on M units of mass for 1 second produce $\dfrac{P}{M}$ units of velocity.

P units of force acting on M units of mass for t seconds produce $\dfrac{Pt}{M}$ units of velocity.

$$\therefore \ V = \frac{Pt}{M}.$$

Provided that we choose our unit of force as above, therefore, we may write

$$MV = Pt,$$

and from this formula we can calculate either the velocity produced in a given mass by a given force acting for a given time :

$$V = Pt/M \ \dots\dots\dots\dots\dots\dots\dots(1),$$

or the force required to produce a given velocity in a given mass in a given time :

$$P = MV/t \ \dots\dots\dots\dots\dots\dots\dots(2).$$

122. A very important particular case of (1) is that in which the force acts for one second ; for the velocity gained in one second measures the *acceleration* produced in the mass M by the force P.

Thus $\qquad\qquad a = P/M, \ \ P = Ma.$

123. The simplest way of solving dynamical problems is to use this formula in conjunction with the Kinematical formulae of the last chapter. Thus :

Dynamical Formula	Kinematical Formulae	
	From rest	Initial velocity u
$a = \dfrac{P}{M}$	$v = at$ $s = \dfrac{at^2}{2}$ $\dfrac{v^2}{2} = as$	$v = u + at$ $s = ut + \dfrac{at^2}{2}$ $\dfrac{v^2}{2} = \dfrac{u^2}{2} + as$

In general, either the forces acting and the mass acted on are given, and it is required to find something about the speed gained or the distance run in a certain time; or else some relation between time, speed, and distance is given, and it is required to find either the force acting or the mass moved, one of the two being known. In the former case we find a from the dynamical formula $a = \dfrac{P}{M}$, and then use its value in the kinematical formulae. In the latter we begin by finding a from the kinematical formulae, and then find P or M, whichever is unknown, from $a = \dfrac{P}{M}$.

124. The unit of Force has been defined, but so far we have no means of comparing it with forces more familiar to us, such as the weight of a pound or of a gramme. This can only be effected by an experiment.

(1) Unit force on the British system of units.

We want to know what force acting on one pound for one second will give it a speed of one foot per second.

Let a standard pound be dropped from a height, so that it is acted on solely by its own weight. At Greenwich it is *found* to acquire in one second a speed of 32·2 feet per second. (§ 69.) The weight of one pound must therefore be 32·2 units of force such as we have chosen; and our unit is the 32·2th part of the weight of one pound at the place where the experiment is tried.

This unit is called the Poundal, or the British Absolute unit of Force, because its value is the same wherever the experiment to determine it is tried. For if it be found that at some other place the speed acquired in one second of fall is different, for instance 32·16 feet per second, then the weight of a pound, as measured by a spring balance, will also be found to be less at this place in the proportion 32·16 to 32·20, so that the 32·16th part of it leads to the same value of the unit.

This value is about a half-ounce weight; and whenever the formulae $a = \dfrac{P}{M}$, or $MV = Pt$ are employed, forces, when given, must be expressed in poundals (by multiplying pound-weights by 32·2); and, when found, will come out in poundals, and can be converted to pound-weights by dividing by 32·2.

(2) The c.g.s. system.

A gramme acted on by its own weight is found at Paris to acquire a speed of 981 centimetres per second, in one second.

The absolute unit of force on the c.g.s. system is therefore the $\frac{1}{981}$th part of the weight of a gramme at Paris. This unit is called a Dyne ($\delta \acute{v} v a \mu \iota \varsigma$ = force). It is not far from the weight of a milligramme. All forces occurring in dynamical formulae must be expressed in dynes by multiplying gramme-weights by 981; and answers expressed in dynes can be converted back to gramme-weights by dividing by 981.

125. Weight is proportional to Mass.

Galileo was led to disbelieve the common opinion of his time that heavy bodies fall more quickly than light ones. To settle the question he tried an experiment at the celebrated Leaning Tower of Pisa. Boxes of the same size and shape, but filled with different materials, were dropped from the summit, and, as he expected, were found to reach the earth at about the same moment, whether their contents were light or heavy. The slight outstanding differences he rightly referred to the resistance of the air, which has a greater proportional effect on the light bodies than on the heavy ones.

Experiment. The student should verify this fact. Let an iron ball, and a wooden ball with an iron nail in it be supported by two small electromagnets attached to a wooden bar and drawn up to any height. If the same current be made to pass round both magnets, the balls can be released simultaneously by breaking the circuit. The balls will reach the floor almost exactly together.

Newton repeated this experiment in a striking form by dropping a guinea and a feather at the same moment in a long glass tube from which the air had been exhausted. The feather fell like the metal.

A very important conclusion follows from this experiment. Since all bodies fall equally fast *in vacuo*, the acceleration must be the same for each at every moment of the fall. Thus $\dfrac{P}{M}$ is the same for all bodies at the same place. But the only force acting on a falling body is its own weight. Therefore W/M is the same for all bodies, *i.e.* the weight of a body is proportional to its mass.

We see now that masses may be compared by comparing their weights, a far more convenient method than that of § 116.

Since the acceleration produced by the weight of a body is the same for all bodies, *i.e.* $g = 32\cdot2$, or 981, according to the system of units employed, we may write

$$W/M = g; \quad \text{or} \quad W = Mg.$$

126. We will consider the experimental evidence for the Second Law after we have discussed the Third Law, meanwhile observing that both of them are abundantly verified by the calculations found in the *Nautical Almanac*. But before leaving the Second Law, we must note two important facts of experience not explicitly stated by it, but implied by its form.

(1) It says nothing about the existing state of motion of the body acted on. The effect of a force in producing new velocity is found to be the same whether the body is at rest or already moving at high speed.

Set two small objects on the edge of a table and sweep a heavy paper-knife along the table so as to strike one of them *horizontally* towards the end of the room, while the other is merely dislodged. They will be heard to strike the floor at the same moment, the one at the foot of the table, the other many feet away.   If a rifle were fired horizontally so as just to dislodge an object at its muzzle, the bullet and the object would still strike the earth at the same moment, though perhaps many hundred feet apart.   Mere speed in no way exempts a body from the action of gravity or any other force.

(2)   The Second Law makes no mention of any other forces that may be acting on the body at the same time.   It is found that every force, however small, produces its whole effect, however great may be the other forces acting on the body.   The attraction of a falling stone upon the earth has its full effect in modifying the earth's motion, although the earth is subject at the same time to the enormous attraction of the sun.   Its motion is compounded of all those produced by the forces acting on it, including the slight pull of the stone.

Newton was the first to state this explicitly in a corollary to the Second Law.   In his tract *Propositiones de Motu*, which preceded the *Principia*, he says:

"Corpus in dato tempore viribus conjunctis eo ferri quo viribus diversis in temporibus aequalibus successive."

If two forces act simultaneously on a body, and if they would respectively produce the motions AB, AC, when acting separately for the same interval of time, then the body will move in that interval to D, since the forces and the motions produced by them are independent of each other.

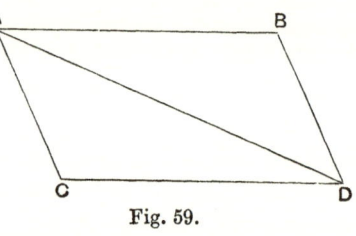

Fig. 59.

In the *Principia* the first Corollary to the Laws of Motion stands thus: "Corpus viribus conjunctis diagonalem parallelogrammi eodem tempore describere, quo latera separatis."

The forces are supposed to be single impulses applied at A, in the directions AB, AC respectively, and sufficient to carry the body to B and C in equal times.   The body will arrive at D, and must have come by the straight line AD, by Law I, for once started, it is not acted on by any force.

In Corollary 2, Newton points out that Statics may be deduced from this principle, and illustrates it by deducing the Principle of the Lever.

This is the first distinct formulation of the Parallelogram of Forces. We shall return to its formal proof later.

127. By the Second Law we can calculate the motion of any body when we know its mass and the forces acting on it. But these forces are applied from without by other bodies. They are pushes, or pulls, or attractions, and there is always a reaction upon the body that pushes or pulls. What is the relation between the forces mutually exerted upon each other by any two bodies? Newton's answer to this question is given in his Third Law.

Law III. *To every Action there is an equal and opposite Reaction.*

Pressure and counter-pressure, action and counter-action are equal. All force is of the nature of a *stress*, that is, a mutual action between two bodies, the same from whichever side it is looked at.

You cannot exert a pressure unless you meet with a resistance, and then the pressure and the resistance grow side by side, being always equal. You cannot cut a piece of paper with one blade of a pair of scissors, nor crack a nut with one half of a pair of nutcrackers. You cannot drive a nail into a board unless it is supported behind, for the board yields before the pressure is great enough to force the nail in. A cannon ball can exert no force till it meets with an obstacle, and then only so great a force as that with which it is resisted. You cannot pull an object harder than it pulls back.

128. Beginners are liable to a difficulty in admitting this. They say: "Why then does any object ever succeed in moving any other at all?" Newton himself considers the case of a horse drawing a cart. No more instructive problem can be found, so we will examine it in some detail.

According to the Third Law the cart pulls the horse backwards as hard as the horse pulls the cart forwards. Everyone will admit that if a spring dynamometer is employed to measure the tension of the traces, it does not matter which end is attached to

the horse, and which to the cart. The reading will be the same.
Then how is it that they ever get into motion ?

The difficulty brings out a very important point. In attacking
any mechanical problem it is essential to begin by fixing upon the
system whose rest or motion we are going to consider. We may
make it include one body, or a collection of bodies, or the whole
universe; but *we must be clear as to what it is.*

In the case of the horse and cart; is it wished to know why
the cart advances ? or why the horse advances ? or why they both
advance ?

Let us begin with the cart. The cart is, then, the system
whose motion is to be determined. It is to be supposed isolated
in our minds from all other objects, as in an imaginary enclosure
acbd.

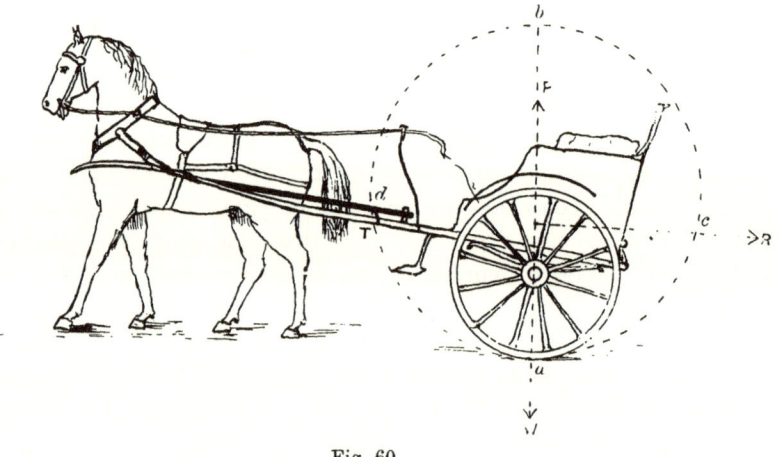

Fig. 60.

Let M be its mass. Let us now go round the enclosure and
see what forces act on it. When we know the forces and the
mass, the Second Law of Motion will tell us what will be its
movements.

(1) The earth exerts two forces on the cart, (*a*) an attraction,
which is the weight W of the cart; and (*b*) an upward pressure P at
the point where the wheel rests on the ground. To each of these,
by Law III, there is an equal and opposite force exerted by the

cart upon the earth. *But with these latter forces we are not concerned at the moment,* since we are not considering the motion of the earth, but of the cart.

The upward pressure (*b*) is also equal to the downward pull of the weight (*a*); for if it were greater than the weight, the cart would rise into the air, and if it were less, the cart would sink into the ground, as in fact it does on soft ground, till the resistance becomes equal to the weight. These forces therefore balance, and may be left out of our account.

(2) The tension of the traces acts forwards. (Observe that there is never any difficulty in deciding which way a tension or pressure acts, if we are clear as to which is the body whose motions we are studying.)

Let the tension be *T*.

(3) The only other forces acting on the cart are certain frictions and resistances, at the axles, on the ground, and against the air. For simplicity, let us suppose that these are equivalent to a direct pull by a rope, backwards, of magnitude *R*.

Now provided the tension *T* be greater than the resistance *R*, there will be a balance of force $T - R$ forwards, and the cart will begin to move, for it will be subject to an acceleration forwards,

$$a_1 = \frac{T - R}{M} \quad\dots\dots\dots\dots\dots\dots\dots(1).$$

Next consider the horse. Let his mass be *m*, and draw an imaginary enclosure round him as before. The forces acting on him through this enclosure are:

(1) His weight, and the upward pressure of the earth, which so long as he stands still will balance as in the case of the cart.

(2) If he tries to go forwards, there will be a tension of the traces pulling him backwards; and this, by the Third Law, will be exactly equal to *T*, the forward tension of the traces upon the cart.

But where is the external force which is required by Law II to make him go forwards? He cannot exert it upon himself. This is implied by Law II, and explicitly stated by Law I. No man can raise himself by pulling at his own boot-straps, or if seated on a chair, by lifting at the lower rails. It must come from outside, from some other object.

Accordingly the horse, besides maintaining the downward pressure necessary to support his weight, *thrusts the earth backwards*; and by Law III the earth immediately thrusts him forwards with an equal reaction. Let this be F. So, too, the skater, gripping the ice with the sharp edge of his skate, gives a back-thrust, and is himself driven forwards; and the swimmer advances by thrusting back a wedge of water from between his legs.

If the horse thrusts back hard enough, F will be greater than T; and he will begin to move with an acceleration

$$a_2 = \frac{F - T}{m} \dots\dots\dots\dots\dots\dots\dots(2).$$

Thus although the back-pull on the horse is the same as the forward pull on the cart, both will advance provided that F is greater than T, and T than R.

Assuming that we know the force with which the horse thrusts, F; the resistances to motion, R; and the masses M, m; we ought to be able to solve the problem completely, and find at what rate motion will ensue. But we have at present only two equations to find three unknowns, viz. the two accelerations, and the tension of the traces. It will be found that there is always a *dynamical* equation for each of the accelerations. If no unknown reaction, such as the tension, comes in, this will be sufficient. But wherever a reaction of this kind occurs, it must be in consequence of some connection between the parts of the system, whereby they have to move in a certain definite relation to each other. That is, for every unknown reaction there will be *a geometrical equation*, expressing the special relation of the parts of the system which gives rise to the reaction. Thus there will always be enough equations to ensure a definite solution.

In the present case, so long as the traces hold, the horse and cart have to advance together, so that the two accelerations are equal. Hence

$$a_1 = a_2 \dots\dots\dots\dots\dots\dots\dots(3).$$

$$\therefore \quad \frac{F - T}{m} = \frac{T - R}{M}.$$

Whence
$$T = \frac{MF + mR}{M + m},$$

and by substitution,

$$a_1 = \frac{T - R}{M} = \frac{\dfrac{MF + mR}{M + m} - R}{M}$$

$$= \frac{F - R}{M + m}.$$

Substitution in (2) would of course give the same value for a_2.

Finally, let us consider the horse and cart as one system. The imaginary enclosure must now be drawn round the two together. The only unbalanced forces acting on the system from without will now be the forward thrust of the earth, F, and the back-pull of the resistances, R. The weights and the upward thrusts of the earth balance as before.

The mass to be acted upon is the total mass of the horse and cart. Hence the acceleration of the whole system will be

$$a_3 = \frac{F - R}{M + m}.$$

The former result can thus be written down directly when a knowledge of the tension is not required.

What has become of the tension in this case? It is no longer an external force, but a mere internal reaction between two parts of the system, which can no more affect its motion as a whole than any of the other forces which hold the parts of the cart and the frame of the horse together. It does not therefore enter into the equations. On the other hand, the tension can only be found by considering the motion of the two parts, between which it occurs, separately.

129. The solution of any dynamical problem by Newton's method involves no principle which has not been illustrated in the elementary question we have been discussing. The rest is mere elaboration of mathematics to meet the complexities introduced by changes in the direction of motion, or of force in curved orbits, or the number of bodies or parts interacting on each other. The process consists of:

(1) writing down the Equations of Motion, *i.e.* equations expressing the accelerations of the different bodies concerned in terms of their masses and the forces acting on them ;

(2) writing down the geometrical equations defining the relative movements of parts which exert reactions on each other ; and

(3) solving the equations.

EXAMPLES.

1. Shew that if a train travelling at 60 miles an hour is suddenly brought to rest by a head-on collision, a passenger facing the engine will strike the opposite wall of the carriage as if he had fallen on to it from the top of a tower 121 feet high.

2. A rain drop experiences a resistance from the air proportional to the square of its velocity.   Hence shew that there is a limit to the velocity it can acquire, and that after this terminal velocity is reached the drop falls with uniform velocity.

3. Find the acceleration :

(1) in foot-second units when

(*a*) a force of 12 poundals acts on a mass of 2 lbs,

(*b*) a force of 14 pounds weight acts on 112 lbs.

(2) in c.g.s. units when

(*a*) a force of 10,000 dynes acts on a mass of 80 gms.

(*b*) a force of 10 pounds weight acts on 20 kilograms.

4. What force (in poundals) is required to produce an acceleration of 12 ft.-sec. units in masses of 2 lbs. ; 12 ozs. ; 112 lbs. ?   Express these forces in pound-weights.

5. How many dynes must act on 2 kilograms to produce an acceleration of 8000 c.g.s. units?   Express this force in gramme-weights.

6. Find the mass acted on when :

(1) a force of 3 pounds weight causes an acceleration of 1, 36, and 96 British units respectively ;

(2) a force equal to the weight of 10 gms. causes an acceleration of 196 c.g.s. units.

7. A force equal to the weight of 3 lbs. acts on a mass of 2 kilograms for 10 seconds.   Find the velocity acquired and the distance travelled, in c.g.s. units.

8. An engine pulls a train of 200 tons mass with a force equal to the weight of a ton and a half. Find the acceleration; the speed acquired (*a*) in half a minute, and (*b*) in half a mile; the distance run in 5 minutes; and the time required to run the first two miles.

9. A train of 120 tons mass running 45 miles an hour is pulled up by the brakes in half a mile. What is the retarding force of the brakes in ton-weights ?

10. A train has its speed reduced from 40 to 20 miles an hour in 400 yards. What is the resistance of the brakes in pounds per ton-weight of the train ? How far and how long will the train continue to run before it stops ?

11. What force is required to stop a train of 100 tons going 30 miles an hour (1) in half a minute, (2) in half a mile ?

12. A one-ounce bullet is fired from a rifle barrel 4 feet long with a speed of 1600 feet per second. What was the average pressure of the powder on the bullet ?

13. A 600 lb. cannon ball is fired from a gun weighing 12 tons with a velocity of 2000 feet per second. If the gun is free to move, with what velocity will it start backwards ?

(By the Third Law the pressure on the gun must throughout the explosion be equal and opposite to that on the shot. Hence the momentum of the gun must be equal and opposite to that of the shot.)

14. A half-ounce bullet in passing through a 2-inch plank has its velocity reduced from 900 to 600 feet per second. Find its acceleration in the plank, and hence the average resistance offered to it. What thickness of plank would have just stopped it ?

15. A curling stone is projected along ice with a velocity of 48 feet per second. Express the resistance due to friction as a decimal of the weight of the stone, if it stops (1) in 300 feet, (2) in 15 seconds.

16. A slinger using a sling 2 ft. 6 in. long whirls a stone weighing 2 ounces in a nearly horizontal circle above his head, so as to make 5 revolutions per second. What is the tension of the string ?
(Calculate first the acceleration of the stone to the centre by § 77.)

17. How many times faster must the earth turn on its axis so that objects at the equator should just lose their weight ?

18. If a train of 200 tons were to run at 60 miles an hour at the equator, what difference would it make in the pressure on the rails if the train were to run due west instead of due east ?

19. A bicyclist is running at 20 miles an hour round a circular track with four laps to the mile. What should be the inclination of his machine to the vertical ?

20. A fly-wheel with a rim weighing 10 tons has a diameter of 6 feet, and runs at 160 revolutions per minute. If there are six spokes, what pull has each spoke to bear in preventing the rim from flying to pieces ?

CHAPTER XIV.

EXPERIMENTAL VERIFICATION OF THE LAWS OF MOTION.
ATWOOD'S MACHINE.

130. THE Laws of Motion may be tested directly by means of Atwood's Machine. This consists of a light pulley of aluminium, balanced very truly on its axis, and supported on a set of friction wheels to reduce the friction as much as possible. A cord passes over it and carries two equal weights, one of which runs in front of a vertical scale. At the zero of the scale is a hinged platform which is pulled flat against the scale by a spring on withdrawing a catch. A clock beating seconds is arranged to withdraw the catch electrically at the instant when the minute hand passes the mark 60″ at the top of the dial.

On the scale are two moveable platforms, the upper one being a hollow ring through which the weight can pass, the lower a solid plate which brings it to rest. Small weights are provided, consisting of flat brass slips which can rest on the top of the main weights, but are too long to pass through the ring-platform.

One of the Atwood Machines, figure 61, is for use in Lecture Demonstrations. It has a scale of 750 centimetres ; and to equalize the weight of thread on each side, an idle thread hangs in a loop from the bottoms of the moving weights. This also gives the experimenter control of the machine from below.

The other Machine, figure 61 *a*, constructed by the Cambridge Scientific Instrument Company, is for accurate work. It has a steel scale, with geometrical contact adjustments for the platforms and fine screw motions for setting them accurately. The times of release and of passing each of the platforms are recorded electrically on a chronograph, and the moving weight is brought to rest in a dashpot at the bottom of the scale column. A weight suspended by a spiral spring vibrates vertically, giving seconds on the chronograph.

131. **First Law of Motion.** When only the two equal large weights are used, there is no unbalanced force tending to set up motion. They will be found to rest in any position, and, if set in motion, they travel uniformly, passing over the same number of centimetres in every second, except for an almost inappreciable loss of speed through friction.

132. **Second Law of Motion.** Place one of the small flat weights across the top of the large weight. Let the masses of the large and small weights be M and m. The only unbalanced force acting on the system is the weight of m.

The total mass to be set in motion is $2M + m$.

Fig. 61.

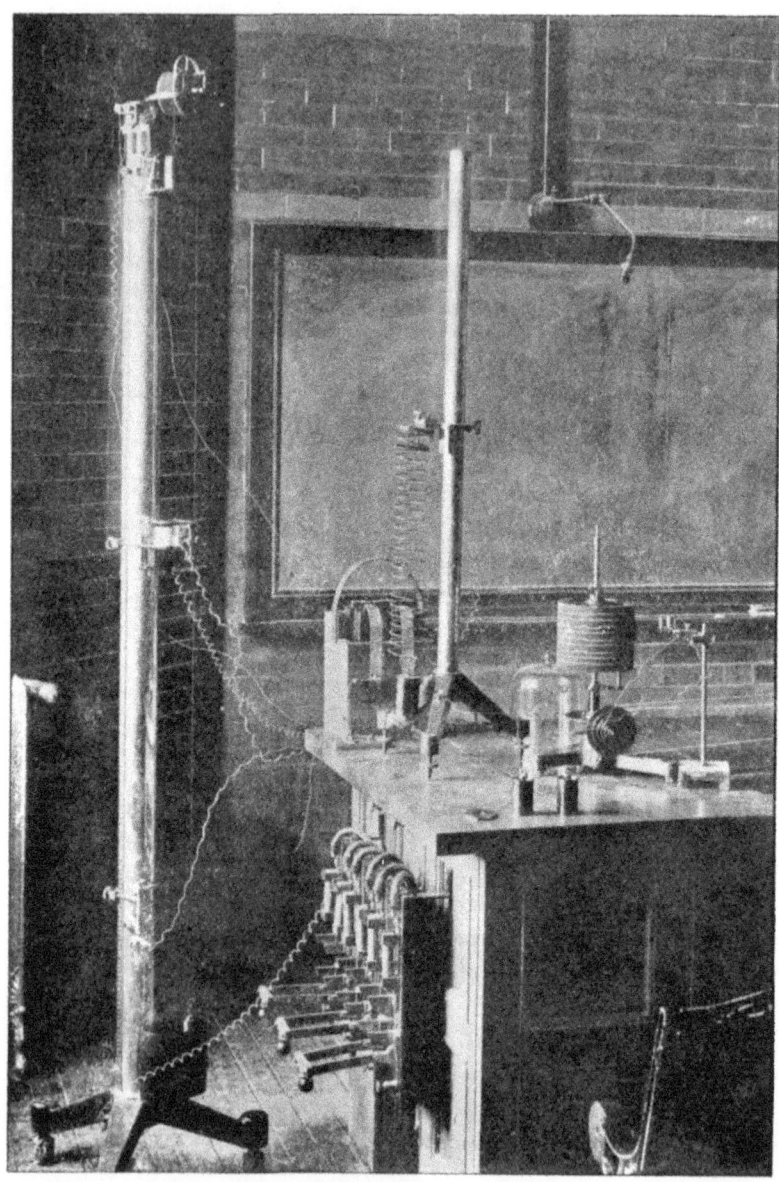

Fig. 61 *a*.

By the Second Law of Motion there should be an acceleration

$$a = \frac{mg}{2M + m}.$$

As this is a constant acceleration, the distance run in t seconds from rest will be $s = \frac{at^2}{2}$, and the velocity acquired $v = at$.

If the weights have been weighed beforehand, a may be calculated, and the values of s and v found for, say, 3 seconds. The ring-platform should be set at the distance s below the zero, and the lower platform near the bottom. Pass the weight carrying the small weight above the top platform; set the platform and lower the weight on to it. On starting the clock the platform will be heard to fall at the beginning of a minute, and the cross weight will be caught by the ring almost exactly at the third following tick. The seconds should then be counted till the last platform is reached.

After passing the ring-platform the system is not acted on by any unbalanced force, so that its velocity is uniform. It may be found by dividing the distance between the two lower platforms by the number of seconds required to pass from the one to the other. By repeated trials the platforms may be set so that the passage of the weights through them coincides with a tick of the clock, and no fractions of a second need be estimated.

Besides moving the large weights and itself, the small weight has to set in motion the aluminium pulley. This can be allowed for by supposing that an extra small mass x must be set moving, so that

$$a = \frac{mg}{2M + m + x} \quad \dots\dots\dots\dots\dots\dots\dots(1).$$

Using a different small weight m', we find a different acceleration,

$$a' = \frac{m'g}{2M + m' + x} \quad \dots\dots\dots\dots\dots\dots\dots(2).$$

a and a' may be determined by experiment with the machine, and x is then found from (1) and (2).

A third experiment should be tried with still another weight m''. The formula with the value of x substituted in it should then give correct results for s and v.

133. The weights and the middle platform are removed from
the machine. The top platform is set, and any
small object placed upon it. The lower platform
is placed at 490 cm., and the clock is started. The
small object is heard to strike the platform exactly at the first tick
after the top platform falls. Its average velocity must be 490 cm.
per second, and therefore the velocity at the end of one second is
980 cm., or 32·2 feet per second.

To find the
Dynamical unit of
Force.

Thus Galileo's experiments at the tower of Pisa may be verified;
and since the object might have been a gramme or pound weight,
the unit force must be the 1/980th part of a gramme weight, or
the 1/32·2th part of a pound weight according to the system of
units adopted.   (§ 124.)

134. The problem of Atwood's machine is the same as that of
the horse and cart (§ 128), with the descending
weights substituted for the horse, the ascending
weight for the cart, and the string for the traces. We have
treated the weights as one system, since only the acceleration was
needed. If the tension T of the string is to be found, the
accelerations a and a' must be calculated separately.   Thus

Third Law of
Motion.

$$a = \frac{(M+m)\,g - T}{M+m} \quad\ldots\ldots\ldots\ldots\ldots\ldots(1),$$

$$a' = \frac{T - Mg}{M} \quad\ldots\ldots\ldots\ldots\ldots\ldots\ldots(2),$$

$$a = a' \ldots\ldots\ldots\ldots\ldots\ldots\ldots\ldots\ldots\ldots(3).$$

Whence

$$T = \frac{2\,(M+m)\,Mg}{2M+m},$$

and

$$a = \frac{mg}{2M+m},$$

as before.

In arriving at this *correct* result, we have assumed that the
string transmits equal and opposite action and reaction T between
the ascending and descending weights, so that the Third Law is
verified.

The advantage of Atwood's Machine is that it enables us to

work with a constant acceleration which may be any small fraction of gravity we choose according to the value of $m/(2M + m)$. It is easier to observe the slower motion, and the two platforms afford convenient means of measuring the acceleration during the first part of the motion, as well as the final velocity acquired.

EXAMPLES.

1. The large weights in an Atwood's Machine are each 223 gms. The rider is 13·4 gms. How many centimetres below the stage must the ring-platform be set so as to be reached in 2 seconds? How far below this must the last platform be set so as to be reached 8 seconds afterwards?

2. With the above weights the ring-platform is set so as to be reached in 3 seconds, when it is found that the lowest platform must be set 429 cms. below it, in order to be reached 5 seconds afterwards. Hence find the value of gravity.

3. In an experiment with the weights in Question 1, it is found that the ring-platform must actually be set at 55·3 cms. below the zero in order to be reached in 2 seconds. On the assumption that the effect of the mass of the pulley, which has also to be set in motion, is equivalent to the permanent addition of a small mass x to the two large weights, an experiment was tried with another rider weighing 27 gms. The platform had then to be set at 108·3 cms. to be reached in 2 seconds. From these two experiments find x and the value of gravity.

4. A mass of 10 lbs. is placed on a smooth table and connected, by a string passing over a smooth pulley at the edge, with a weight of 2 lbs. hanging freely. Find the acceleration of the system. How far from the edge must the 10 lb. weight be placed so as to fall off in 2 seconds?

5. Find the tension of the string in Question 1 (before and after the ring-platform is reached) and in Question 4.

CHAPTER XV.

135. The Newtonian method of solving dynamical problems begins by fixing the rate at which the velocity changes, *i.e.* the acceleration, by means of the formula $a = P/M$.

From this it proceeds to calculate what will be the velocity, and where the body will be at every succeeding instant, by the kinematical formulae:

$$v = at \text{ and } s = at^2/2,$$

or their proper modifications if the acceleration is variable.

136. For many purposes we do not need to follow the motion with such intimate scrutiny, but are content to fix our attention on the final velocity acquired when some other position is reached, disregarding what has gone on meanwhile. This we can do by means of the third kinematic formula

$$v^2/2 = as,$$

or its dynamical equivalent obtained by multiplying both sides by the mass M. Thus

$$Mv^2/2 = Mas = P \times s \quad \dots\dots\dots\dots\dots(1).$$

This formula tells us what speed will be acquired by a mass M, when a force P pushes it steadily through a distance s; or, if the mass be already in motion with given velocity v, and the force P be employed to stop it, it tells us how far the force will be driven back, before the body is brought to rest, viz.

$$s = Mv^2/2P.$$

The product $P \times s$ on the right-hand side of (1) has already received a name. It is the work done by the force P while the body travels the distance s. (§ 45.)

137. A body or system on which work has been done is found to have an increased power of doing work itself, that is, of producing physical changes in other bodies. It is therefore said to possess more energy than before. This energy may take two forms.

138. The body may have been set in motion with regard to
I. **Energy of Motion.** other bodies. By the first law of motion it will then continue in motion until some resisting force is employed to stop it. Let it be brought to rest by the constant force P in a distance s. Before it stops, it does a quantity of work $P \times s$, and since

$$Mv^2/2 = P \times s,$$

it appears that the work done by a body of mass M and velocity v before it can be brought to rest is measured by $Mv^2/2$.

The quantity of work which the moving body can do in virtue of its motion is called its *Kinetic Energy*, or energy of motion, and is measured by $Mv^2/2$.

139. That motion confers on a body a certain "efficacy" or
Comparison of Energy and Momentum. power of overcoming opposing forces, and producing changes in other bodies was of course a familiar fact. For a long time a controversy raged between the followers of Des Cartes and Leibnitz as to whether this "efficacy" of the body was proportional to the velocity or to the square of the velocity. The dispute is seen to be meaningless if we compare the Newtonian formula,

$$MV \quad = \quad Pt,$$
$$\text{(Momentum)} \quad \text{(Impulse)}$$

with the energy formula, first employed by Huyghens,

$$\frac{Mv^2}{2} \quad = \quad Ps.$$
$$\text{(Energy)} \quad \text{(Work done)}$$

Given a mass M moving with a velocity V. The first formula tells us *how long* ($t = MV/P$), the second *how far* ($s = MV^2/2P$) it will continue to move against a given force P, before it can be brought to rest.

Conversely the two formulae respectively give the time and the distance required by the force P to produce in M, initially at rest, the velocity V.

Thus both the Cartesians and the Leibnitzians were right. The time during which a moving body will go on overcoming a given resistance is proportional to its velocity, the distance to the square of its velocity. A body thrown vertically upwards, and then again with twice the speed, will in the second case rise for *twice* as long a time, but *four times* as high as before.

A moving body is just a moving body, and its effects can be calculated when we know its mass and its velocity. In making the calculations we often have to reckon the product Mv; often again the product $Mv^2/2$; so often, in fact, that it is worth while to denote them by the special names Momentum, and Energy. But it must not be supposed that these are occult properties of the body, which may sometimes exhibit one, and sometimes the other. Every moving body has both Momentum and Energy, *i.e.* we can calculate both MV and $MV^2/2$ for it. Which of the two we should choose, depends on the purpose we have in view.

140. Let us compare in respect of momentum and energy an ironclad of ten thousand tons mass, so nearly at rest that it is moving only one inch per second, with a one-ounce bullet moving 1600 feet per second.

(1) Momentum $= MV$. This is

For the ironclad $2240 \times 10,000 \times 1/12 = 1,866,666 \cdot 6$ British units
,, bullet $1/16 \times 1600$ $= 100$,, ,,

(2) Energy $= MV^2/2$.

For the ironclad $\dfrac{10,000 \times 2240}{2} \times (1/12)^2 = 77,777 \cdot 7$ foot-poundals.

,, bullet $\dfrac{1}{16} \times \dfrac{(1600)^2}{2}$ $= 80,000$,, ,,

The ironclad has enormously greater momentum than the bullet; but the bullet has rather the greater energy.

To understand precisely what this means, let us suppose that each has to be brought to rest by holding against it a perfectly hard shield with a steady force of, say, 1000 poundals. We may

ask; (1) How long will each go on pushing the shield back?
Or (2) How far?

(1) Since $MV = Pt$; therefore $t = \dfrac{\text{momentum}}{\text{force}}$,

For the ironclad $t = \dfrac{1,866,666\cdot\dot{6}}{1000} = 1,866\cdot\dot{6}$ seconds.

„ „ bullet $\quad t = \dfrac{100}{1000} \qquad = 1/10$ second.

(2) Since $MV^2/2 = P \cdot s$; therefore $s = \dfrac{\text{energy}}{\text{force}}$,

For the ironclad $s = \dfrac{77,777\cdot\dot{7}}{1000} = 77\cdot\dot{7}$ feet.

„ „ bullet $\quad s = \dfrac{80,000}{1000} = 80$ feet.

The bullet will be brought to rest in the tenth part of
a second, whereas the ironclad will creep on for more than half
an hour. Nevertheless the bullet will push the shield back
through 80 feet, as against 77·7 feet for the ironclad.

This is easily understood. For the average speed of the
bullet, which decreases uniformly from 1600 feet per second to
nothing, is 800 feet per second. That of the ironclad is half an
inch per second. The bullet in one-tenth of a second travels its
80 feet, while the ironclad only creeps over 77·7 feet in its half
hour.

141. Although a body on which work has been done may be
II. Energy of at rest, there may have been a change in its
Position. position with regard to other bodies; or in its
shape, that is, the position of its own parts with regard to each
other. And in consequence of this change it may be capable of
doing more work than before, *i.e.* possess more energy.

To raise a one-pound clock-weight through a height of three
feet, three foot-pounds of work must be done. The weight has
then been *drawn apart a distance of three feet from the earth
against the attraction of gravity.* The " system," consisting of the
earth and the weight separated from it, has the power of doing
this work back again; for, if allowed to run down, the weight will
do so, and drive the clock for a long time. The three foot-pounds

c. 10

of work have thus been stored up in the system, which possesses three foot-pounds of energy more than it had before. This is often called the energy of the clock-weight, but it should never be forgotten that the energy really resides in the system, earth-and-clockweight, and is due to the relative separation of its parts against the force of gravity acting between them.

Similarly, the water in a lake or mill-dam can do work if allowed to fall to a lower level; to the amount of one foot-pound for every pound of water that descends a vertical height of one foot.

Again, work must be done to wind up a watch-spring, or bend a bow. The coiled watch-spring and the bent bow possess an equivalent amount of energy, and will do the work back again if allowed to unbend. In these cases the particles on the inner side of the spring or bow have been forced together, and those on the outer side drawn apart, against the elastic forces of cohesion that hold the bow together. When the constraint is released, these forces bring the particles of the bow back to their positions of equilibrium, and thus do upon the arrow the work that was stored up in the bow by the force used in bending it. The arrow leaves the bow with the corresponding amount of kinetic energy, which, again, it can give up only on meeting some obstacle, or otherwise experiencing a force from the action of some body not moving at the same rate.

142. The name *Potential Energy* has been given to that form of energy, or capacity for doing work, which is due to a mutual displacement of objects or parts of an object against the forces which hold them together. It is quite possible that all forms of energy, including those now classed as Potential, may ultimately be reduced to cases of kinetic energy. Meanwhile the word Potential must not be misunderstood to mean that this form of energy is not really energy at all, but only something which may become energy, if allowed to convert itself into the kinetic form.

143. No formula can be given for the calculation of Potential energy, so universally applicable as the expression $MV^2/2$ for the energy of a moving body. But very often it can be reckoned easily. The potential energy of raised weights is at once expressed in foot-pounds by multiplying the number of pounds by the height in feet through which they are to fall. In the case of elastic bodies, we often know the force which was employed in producing the distortion. Thus if we know the average force used in drawing the bow, and the length of the arrow drawn, the product gives the energy stored in the bent bow.

144. Wherever bodies are in motion under the action of forces, work is being done, and equivalent amounts of energy stored in the system or expended. Mechanical processes, indeed, may be regarded as cases of the transfer of energy from one body or system to another, or transformations from one of its forms to another. Every such transfer involves the exertion of a force while the body moves over a certain distance. We may at our pleasure either (1) regard the transfer of the energy as the result of the work done by the force, and measure it by the product of the force into the distance moved, $P \times s$; or (2) consider the force a manifestation of the transfer of energy. From this point of view the force is measured by the quotient of the energy transferred, or work done, divided by the distance moved, *i.e.* the force is the *space-rate* of transfer of energy.

(1) In drawing a bow the archer exerts a force on the arrow and draws it through its length. He does

Illustrations of the Transfer of Energy.

work, which is stored as potential energy in the bent bow, in virtue of the relative displacement of its parts. When the arrow is released, the bow does this work on the arrow, exerting a force on it by means of the string, while the arrow moves its length. The energy is thus transferred to the arrow, and at the same time transformed from potential to kinetic energy, due to the motion of the arrow relative to the earth, including the bow and other surrounding objects. The energy cannot be transferred again until the arrow meets some object, relatively at rest, which can exert a force upon it; but if the arrow is rising against gravity, part of its kinetic energy will be transformed into potential energy; so that at every moment of its flight it possesses, in place of the kinetic energy it has lost, an equivalent amount of potential energy, due to its separation from the earth against its weight.

(2) In the common pendulum the transformation of energy from the potential to the kinetic form, and back again, takes place at every swing. The energy is first stored in the pendulum, when it is drawn aside by the exertion of force, as potential energy measured by the product of the weight of the bob into the vertical height above the lowest position. During the descent the potential energy is gradually converted into equivalent energy of motion, and has become entirely kinetic at the bottom of the swing. At each point of the descent the sum of the kinetic and potential energies is the same. During the ascent the kinetic energy is expended as the pendulum climbs against gravity, and when it reaches the same height on the other side, the whole of it has again been converted into potential energy, and the swing recommences.

This process might be repeated indefinitely but for frictions and resistances, hitherto left out of consideration, which absorb a small quantity of energy in each swing. It is the business of the descending clock-weight to supply this small loss by means of the escapement, and so maintain the swing. The energy of the clock-weight, again, is supplied by the work done in winding it at intervals.

(3) A similar instance in which it is easier to trace the force exerted during the transfer of energy, is found in Atwood's Machine (§ 131), when the two equal weights A, B are allowed to run alone, and therefore uniformly. No change is taking place in the kinetic energy of either weight. But the potential energy of the descending weight A, with regard to the earth, is decreasing; and that of the ascending weight increasing at the same rate. The force effecting the transfer is the tension of the string, which is doing work against the one, and an equal amount of work upon the other.

145. Many mechanical processes depend upon the storing of
Storage of   energy in some system, by doing work upon it,
Energy.   and then allowing the system to give up the work
either suddenly or slowly according to convenience.

Thus the clock-weight or watch-spring is wound up in a few seconds, and gives out its stored energy slowly, during eight days or twenty-four hours. The bow gives out its energy almost instantaneously, exerting the same force on the arrow as that used by the archer in drawing it, but following up the arrow, as its speed increases, far more swiftly than the archer could have done, besides enabling him to do the hard work at leisure, and then concentrate his attention on the aim.

The Hammer enables us to exert a much greater force than we could unaided. Let us suppose that a hammer-head weighs one pound, and that we draw it down by the handle with a force of ten pounds-weight, through a vertical height of two feet, on to the head of a nail. What must have been the pressure exerted on the nail, if we find it is driven in half an inch by the blow?

During the descent both the pull of gravity (1 pound) and the force (10 pounds) have been doing work, so that energy has been stored in the hammer-head to the extent of

$$11 \times 2 \text{ foot-pounds.}$$

It is in the form of kinetic energy at the moment of striking; and assuming that it is all taken up in overcoming the resistance of the nail (average value $= R$), while the nail recedes half an inch ($= \frac{1}{24}$ ft.), we have

$$R \times \tfrac{1}{24} = 11 \times 2.$$
$$\therefore \quad R = 528 \text{ pounds-weight.}$$

This is the resistance required to exhaust the kinetic energy of the hammer in half an inch. But the weight of the head continues to act during the process, and if the force of 10 pounds is also applied, we must add 11 pounds more, so that the total average pressure must have been 539 pounds. This is the dead weight which the nail could just support without being driven in.

The Punching Machine is a machine for punching holes through thick plates of metal. At first sight it is difficult to conceive of any tool being driven with such force as to cut a $\frac{3}{4}$-inch hole through an inch plate of cold steel quickly and quietly. It is easily done as follows. The tool is attached to the short end of a lever whose long end is forced up by a cam, or projection on a wheel, which only comes round once in every seven or eight revolutions of the engine which drives the machine. During six or seven strokes the engine does work on a heavy fly-wheel, and when the cam comes round, the whole of the kinetic energy stored in the now rapidly revolving wheel is brought to bear on the tool. When it comes in contact with the plate one of three things must happen. Either (1) this energy must disappear, the machine being suddenly brought to rest without equivalent work done; or (2) the machine must break; or (3) the plate must be punched. But the laws of motion will certainly not fail; and it is the business of the manufacturer to make the machine strong enough not to break. The only alternative is that the plate must be punched; and accordingly it is.

146. Heavy fly-wheels are used for another purpose, to ensure the steady running of an engine. The steam does work on the piston at very different rates at different parts of the stroke; and at the beginning and end of the stroke, the two dead points, no work is being done at all. If the engine were coupled directly to the machinery of a factory, each machine would run in a series of jerks; and should one or two machines be disconnected or brought into action, the speed of all the rest would be suddenly and seriously affected. But if a very heavy fly-wheel be attached to the shaft, the engine pumps energy into it at a rate varying throughout the stroke, but the machines draw off their supply from the large store accumulated in the wheel. The energy of such a wheel can be calculated (§ 260) when its mass, dimensions, and speed are known. It is then easy to design a wheel whose energy at the normal speed shall be any number of times the whole amount supplied by the engine during one stroke; so that the speed cannot vary during the stroke by more than a small fraction, and will not change very greatly even if a number of extra machines be suddenly thrown in or out of gear. The water supply of a large town is managed on precisely the same principle. The engines are not connected directly to the service pipes, or the water would issue in sudden jets; but the pumps work into a large reservoir, from which the supply is drawn off. No appreciable change is caused in the level of the reservoir (and therefore in the steady pressure of the service) either by the

Regulation of the Supply of Energy.

intermittent strokes of the pumps, or by the casual turning on or off of taps in other parts of the city.

147.　In all the cases so far considered there has been no gain or loss of energy on the whole, but only a transfer from system to system, or from one form to another.　What has been gained or lost in one shape has been lost or gained in another.　The work done in winding up the clock-weight can be recovered by letting it run down; the bent bow restores in unbending the work required to bend it; the pendulum rises to an equal height on the other side of the vertical.

Reversible and Irreversible Processes.

Mechanical processes of this kind, which can be run backwards with recovery of the whole of the original work, are called *reversible.*

In practice they are generally accompanied by others which are irreversible.　The Simple Machines and their combinations do not give the results demanded by the formulae we have obtained.　Their movements are interfered with by frictions and resistances, wherever their moving parts come into contact with each other or with the surrounding air.　Even the pendulum is affected by resistances at the pivot and against the air.

These forces, being only called forth by motion, by their very nature always act so as to oppose it.　When therefore the machine is run backwards, their direction is reversed, and instead of the work expended against them being restored, more is used up. It is the business of Physics to trace what becomes of the energy which thus passes away from the ken of Mechanics.　Here it may only be said that when account is taken of all the other effects accompanying mechanical processes,—the heat, the sounds, the luminous, electric, magnetic, and chemical changes,—it is found that the total energy of a system, isolated and left entirely to itself, though it may take on many forms, is unalterable in amount. This is the doctrine of the Conservation of Energy, the central landmark of the science of the nineteenth century.　The principle began to be clearly apprehended from 1840 onwards, first in the case of Heat through the work of Joule and Mayer; and was first formally extended to all branches of Physics in 1847 by Helmholtz in his paper on " Die Erhaltung der Kraft."　Maxwell gives it the following general statement:

" The total energy of any body or system of bodies is a quantity which can neither be increased nor diminished by any mutual action of these bodies, though it may be transformed into any of the forms of which energy is susceptible."

148. The Mechanical Powers and their combinations are incapable of producing a supply of work. They can only transfer or transform an existing supply, so as to apply the force in some specially convenient way. They may increase the force, but in this case the distance through which it moves the body is decreased in the same proportion, or decrease the force, but in this case the distance is increased in the same proportion. *" What is gained in power is lost in speed."* Machinery cannot produce work for us; it has to be worked.

Sources of Energy. The Measurement of Power.

The energy needed for driving a machine must be obtained from such sources as living animals, the kinetic energy of winds, the potential energy of water-falls, the chemical energy stored in coal or other fuel. From these must be supplied not only the useful work delivered by the machine, but the waste and loss due to irreversible processes in working it.

149. The Power of any of these agents is measured by the rate at which it can supply work. Many years ago James Watt made experiments on the power of some of the heavy dray-horses belonging to Barclay and Perkins' Brewery, London. The horses were set to raise a weight of 100 lbs. from the bottom of a deep well by pulling horizontally on a rope passing over a pulley. Watt found that a horse could walk about $2\frac{1}{2}$ miles an hour at this work, thus doing $2 \cdot 5 \times 5280 \times \frac{100}{60} = 22,000$ foot-pounds per minute. Allowing 50 °/$_0$ extra for the work wasted in frictions, he arrived at the estimate of 33,000 foot-pounds per minute for the average power of a horse. This unit has been adopted by engineers, and is known as a *Horse-Power*.

150. When several forces act on a moving body, each does work, or has work done against it. If the work done by some of them is equal to the work done against the others, the kinetic energy of the body is unchanged.

The case of Equilibrium.

But uniform motion can only take place when on the whole no forces are acting. The forces in this case must therefore balance, or be in equilibrium. This is, in fact, the Principle of Virtual Work, already treated in Chapter VII.

Thus when the equal weights of an Atwood's Machine (§ 131) are running uniformly, the upward tension of the string on either weight is exactly equal and opposite to the downward pull of gravity. The forces are in equilibrium, just as they would be in the special case when the velocity of the system was zero, *i.e.* when it was at rest relatively to the earth.

We shall now consider the equilibrium of forces from the Newtonian point of view in more detail.

EXAMPLES.

(*Caution.* In using the dynamical formulae $MV = Pt$; $MV^2/2 = Ps$, the force must always be expressed in the proper dynamical units, *i.e.* poundals or dynes, according to the system employed ; conversely, forces determined from these formulae must be converted to pound- or gramme-weights by dividing by 32·2 or 981 respectively.)

1. Calculate in foot-pounds the energy of :

    (*a*) a projectile weighing 1034 lbs., and having a velocity of 2262 ft. per second ;

    (*b*) a train of 300 tons moving at 60 miles an hour.

2. The projectile in Question 1 penetrates a sandbank to a depth of 30 ft. : the train is brought to rest by the brakes in one minute. Compare the resistance offered by the sandbank with the retarding force of the brakes.

3. The unit of Power commonly employed in electrical engineering is that of the c.g.s. system, the Watt, which is a power capable of doing 10^7 ergs of work per second. If 1 lb. = 454 gms., 1 metre = 39·37 inches, and $g = 981$, shew that one horse-power = 746 watts.

4. A man weighing 12 stone climbs a mountain at the rate of 1000 feet (vertically) an hour. What horse-power is he developing?

5. What must be the horse-power of an engine to pump 1100 cubic feet of water per hour from a well 120 feet deep, a cubic foot of water weighing 1000 ounces ?

6. A train weighs 120 tons including the engine. The resistances to motion on a level are equivalent to a retarding force of 16 lbs. weight per

ton. Find the greatest speed at which the train can run if the engine is of 150 H.P.

("Full speed" is the speed at which the engine is just able to exert a force equal to the resistances to motion. The train then moves uniformly under the First Law of Motion.)

7. If the train in Question 6 is moving at 20 miles an hour, and the engine is working at full power, find the acceleration.

8. Find the H.P. of an engine which can take a train of 100 tons up an incline of 1 in 200 at 20 miles an hour, the resistances being equivalent to 14 lbs. per ton. (Here the engine has to do the work required to lift the train vertically through a certain height per minute, as well as to overcome the resistances.)

9. The resistances to motion of a train being 14 lbs. per ton (English) weight, if the train going 40 miles per hour come to the foot of an incline of 1 in 168, the steam being turned off, find how far it will run up the incline.

If it had come to the top of the incline, how far would it have descended before stopping?

10. A bullet weighing half an ounce is fired with a speed of 2000 feet per second from a rifle weighing 10 lbs. If the rifle kicks back through 3 inches, find the average pressure applied by the shoulder in bringing it to rest.

11. A hammer-head weighing 1 lb. strikes a nail with a velocity of 10 feet per second, and drives it in 1 inch. What was the average pressure of the hammer on the nail?

12. The ram of a pile-driver weighs 200 lbs. and falls 12 feet on the head of a pile which yields half an inch. What steady weight could the pile sustain?

13. A clock-weight of 4 kilogrammes is wound up through a height of 1 metre and then drives the clock for eight days. Express in Watts the power needed to drive the clock.

14. In a steam engine the average pressure of steam during the stroke is 180 lbs. on the square inch. The length of stroke is 3 ft. 4 in., and the diameter of the piston is $5\frac{1}{2}$ inches. If the engine makes 125 revolutions per minute, find its horse-power.

15. An ocean steamer with engines of 30,000 H.P. can make 25 miles an hour. What is the resistance to her motion through the water?

16. The average flow over Niagara Falls is 270,000 cubic feet per second. The height of fall is 161 feet. What horse-power could be developed from the Falls if all the energy were utilized?

17. A belt is transmitting 12 H.P. to a pulley 2 feet in diameter, running at 375 revolutions per minute. What is the driving force of the belt?

CHAPTER XVI.

THE PARALLELOGRAM LAW.

151. ACCORDING to the Second Law of Motion, when two or more forces act at a point of a body, each produces its effect independently of the others, and this effect is not only proportional to the magnitude, but takes place in the direction of the force.

Forces may therefore conveniently be represented by straight lines. For a force is completely specified when

Representation of Forces by Straight Lines.

we know (1) the point at which it acts, (2) its direction, and (3) its magnitude; and a straight line can be drawn (1) from any point, (2) in any direction, and (3) of such a length as to represent any magnitude on any convenient scale.

Quantities which, like forces, depend for their effect on their direction as well as on their magnitude, are distinguished as *vector* quantities, while quantities which have only magnitude, such as a sum of money, or the amount of corn in a heap, are called *scalar* quantities. Scalar quantities are added by ordinary arithmetic. But a special rule is required for adding, or rather compounding, vector quantities. This rule is the Parallelogram Law, already stated for forces in § 38. We proceed to prove it in turn for Displacements, Velocities, Accelerations, and Forces.

152. Let a point O receive two separate displacements represented by OA, OB respectively. The order in which the displacements are given is immaterial. We may suppose the point first carried to A, and then displaced through AC, equal and parallel to OB; or first carried to B,

I. Displacements.

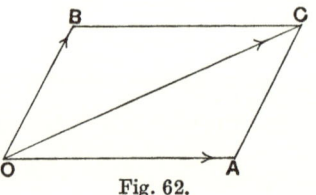

Fig. 62.

and then displaced through BC, equal and parallel to OA. The joint result is the same. The point arrives at C, which it might have reached by a single displacement represented by OC.

A single displacement can thus be found which is *equivalent to* (*i.e.* has the same effect as) any two displacements; it is represented by that diagonal of the parallelogram constructed on the lines representing the displacements which passes through the point.

153. If the two displacements OA, OB, take place uniformly and simultaneously in the course of one second, OA, OB will represent velocities, and OC the single velocity which is equivalent to them. The rule is thus true for velocities.

II. Velocities.

If any difficulty is found in conceiving that a point may have two velocities at once, think of a fly crawling along the paper along OA in one second, while the paper itself is moved obliquely along OB. The velocity of the fly with regard to the table is OC, which may be regarded as made up of his velocity with regard to the paper, OA, together with that of the paper with regard to the table, OB.

If the motions take place uniformly, by the time the fly reaches any point A', the paper will have moved a proportional distance $A'C' = OB'$ such that

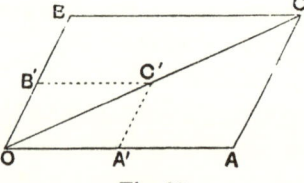

$$\frac{A'C'}{OA'} = \frac{AC}{OA},$$

so that C' is on OC, and, in one second, he actually moves along OC, relatively to the table.

Fig. 63.

154. (1) A steamer steering due East at 10 knots an hour is carried by a current due North at 3 knots an hour. Find the real speed and course.

Applications.

The speed (Fig. 64) is

$$OC = \sqrt{10^2 + 3^2} = \sqrt{109} = 10\cdot44,$$

$$\tan\theta = \tfrac{3}{10},$$

$$\theta = 16^\circ\ 42'\ \text{north of East.}$$

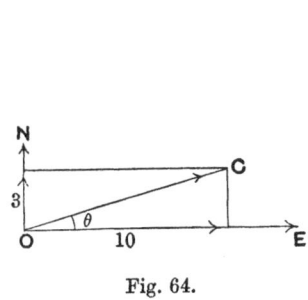

Fig. 64.

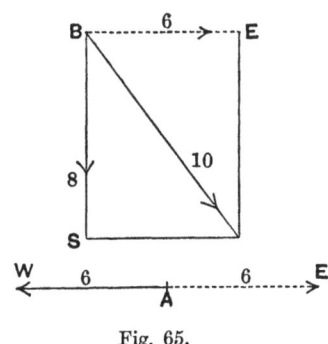

Fig. 65.

(2) A vessel makes 6 knots an hour due West. Another is making 8 knots due South. What is the speed and course of the second with regard to the first ?

Cases of relative motion, such as this, are best solved by the following artifice. No difference will be produced in the *relative motion*, if each of the moving objects is given an extra velocity, provided it is the same for each.

Let A, B (Fig. 65) be the vessels. Apply to each the velocity 6 knots due East, which is equal and opposite to the actual velocity of A. The effect will be that A is reduced to rest, having equal and opposite velocities; while B moves with the two speeds 8 knots South and 6 knots East jointly. But these are equivalent to a speed $\sqrt{6^2 + 8^2} = 10$ knots, at an angle $\tan^{-1} 3/4$ east of South. This is the speed and course relative to A supposed at rest.

If the position of B with regard to A is given, it is easy to calculate whether there will be a collision, or what will be the shortest distance between the ships.

155. If the velocities OA, OB are communicated to a point
III. Accelerations. during one second, it has accelerations OA, OB;
but it is obvious that the effect is the same as if
the equivalent velocity OC were imparted every second. That is,
an acceleration OC is equivalent to the two accelerations OA, OB.

156. Let two forces represented by OA, OB act on the
IV. Forces. same mass. By the
Second Law of Motion
they will produce, independently, accelerations in the directions OA, OB
and proportional to them. OA, OB
may therefore be taken to represent
the accelerations.

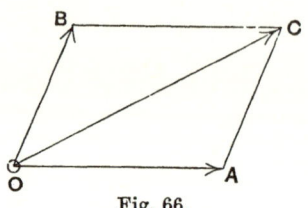

Fig. 66.

But a single acceleration OC is equivalent to OA, OB jointly;
and by the second law this might have been produced by a single
force acting in the direction OC, and represented by OC on the
same scale as that on which the original forces are represented
by OA, OB.

Hence the force OC is equivalent to the two forces OA, OB,
and the rule is true for forces.

EXAMPLES.

1. A train is travelling due North at 20 miles an hour through a shower
of rain falling almost vertically, but with a slight inclination eastwards,
enough to make the drops graze the windows. If the raindrops have a speed
of 16 feet per second, find the inclination to the vertical of the splashes on
the windows.

2. A shot with a velocity of 2000 feet per second is fired at a steamer in
a direction at right angles to the steamer's course, and pierces both sides.
If the deck is 40 feet broad, and the steamer is making 25 miles an hour,
find how many inches the second hole will be astern of the first.

3. The speed of the earth in her orbit is 19 miles per second. Consequently the light from a star appears to be slightly altered in direction to
an observer on the earth, and the star is apparently displaced in the direction
of the earth's motion. This "aberration" from the true position (discovered
by Bradley in 1729) is 20·45″ for a star situated in a direction perpendicular
to the earth's line of motion. Hence find the velocity of light.

CHAPTER XVII.

THE COMPOSITION AND RESOLUTION OF FORCES.
RESULTANT. COMPONENT. EQUILIBRIUM.

157. By means of the Parallelogram of Forces two or more forces acting at a point may be compounded into a single force, called their *Resultant*, which shall produce the same effect. And this effect can in general be more easily calculated for the single resultant than for the several forces to which it is equivalent.

In particular, when there is to be no change at all in the state of rest or motion, *i.e.* when there is to be equilibrium, the resultant must, by the Second Law of Motion, be zero. Any mathematical expression of this fact is a statement of the *conditions of equilibrium* for the given forces.

Conversely, a single force may be resolved into two or more *Components*, which shall together have the same effect. This is often convenient especially when we wish to limit our attention to the motion or conditions of equilibrium in a particular direction; for each force can be resolved into a component along that direction, and another perpendicular to it; and the latter may be disregarded, as it can produce no effect at right angles to its own line of action.

The magnitudes and directions of the straight lines representing the Resultants and Components are to be found by geometrical construction or trigonometrical computation.

158. I. To find the Resultant of Two Forces.

Forces acting in the same plane at the same point.

(a)   Geometrical Methods.

(1)   Construct the parallelogram and draw the diagonal.

(2)   The whole parallelogram need not be drawn. Take OA to represent the first force, and from A draw AC representing the second in magnitude and direction (but not in point of application).

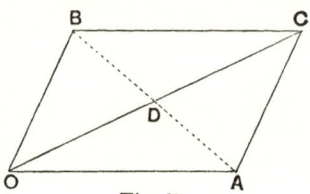

Join OC. This is the Resultant.

It is better, especially when there are several forces, to make two figures:

Fig. 67.

a force diagram, where the lines represent the forces completely, and a construction diagram of the triangles giving the magnitudes and directions. Thus :

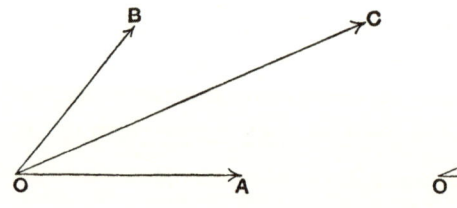
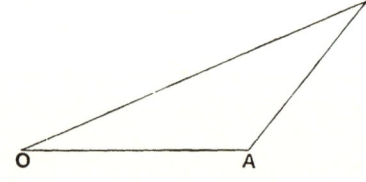

Fig. 68.

(3)   Since the diagonals bisect each other, $OC = 2OD$, where D is the middle point of AB. This value of the resultant is occasionally convenient.

(b)   Trigonometrical Method.

Let the two forces be P and Q, inclined at an angle α; let R be their resultant, making an angle θ with P.

Then $OC^2 = OA^2 + AC^2 - 2OA \cdot AC \cos OAC,$

$$\therefore R^2 = P^2 + Q^2 - 2PQ \cos(180° - \alpha)$$

$$= P^2 + Q^2 + 2PQ \cos \alpha.$$

And $\dfrac{\sin \theta}{Q} = \dfrac{\sin A}{R},$

$$\therefore \sin \theta = \frac{Q}{R} \sin (180° - \alpha)$$

$$= \frac{Q}{R} \sin \alpha.$$

(*Caution.* If other forces have to be combined with the resultant of these two, the whole work has to be done over again for each force, and the expressions become very cumbrous. The method must never be employed for three or more forces, though it is occasionally convenient when there are only two. For more than two the method of § 164 must be used.)

The special case when the forces are at right angles is important.

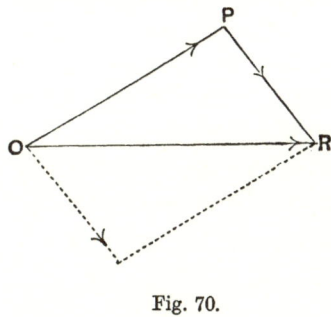

Fig. 69.

Here $R^2 = P^2 + Q^2,$

and $\tan \theta = \dfrac{Q}{P}.$

159. II.   To resolve a Force into two Components.

This can be done in an infinite number of ways. For draw any triangle on the line representing the force (Fig. 70). The sides give the magnitudes and directions of two components equivalent to the force. Thus:

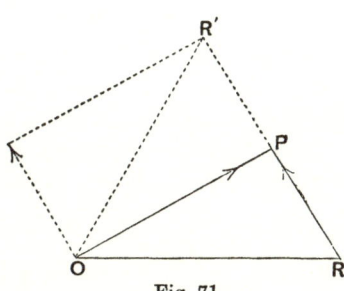

Fig. 70.   Fig. 71.

Observe that the sides of the triangle must be *taken in order*, *i.e.* we must continue along them in the same direction round the triangle. The resultant of OP and RP, applied at O, is not OR but OR' (Fig. 71).

160. (1) To resolve a force into two components of given magnitudes.

This is to construct a triangle when the three sides are given (*Euc.* I. 22). Evidently the two components must together be greater than the force, or there is no solution.

161. (2) To resolve a force into two components in given directions.

Let OR represent the force; and let Ox, Oy be the directions.

Draw parallels to the directions through R. OP, OQ are the components (Fig. 72).

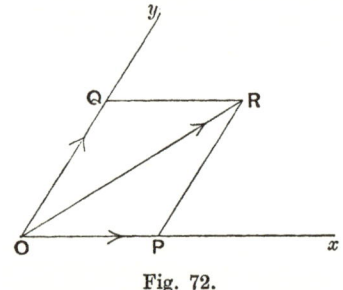

Fig. 72.

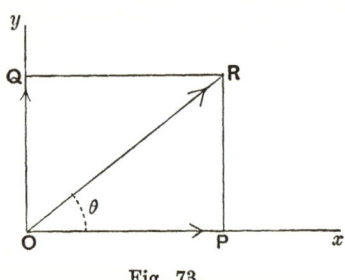

Fig. 73.

162. Let the force OR and the direction Ox be fixed, while Oy varies. Then for every different direction given to Oy the component OP along Ox has a different value. The special case when OQ is at right angles to OP is so important that the value of the component OP in that case is called *The Resolved Part* of OR in the direction Ox, the corresponding value of OQ being *The Resolved Part* of OR in the perpendicular direction.

Let X, Y be the resolved parts of the force R, represented by OR (Fig. 73), along Ox, Oy, and let R make the angle θ with Ox.

Then $\qquad X = OP = OR \cos \theta = R \cos \theta,$

and $\qquad Y = OQ = OR \sin \theta = R \sin \theta.$

To find the resolved part of a force in any direction multiply it by the cosine of the angle between the force and that direction.

Since $\sin \theta = \cos (90° - \theta) = \cos QOR$, the rule just stated applies to the component Y as well as to the component X. If the component in any direction is found by multiplying by the

cosine or sine of any angle, then the component in the perpendicular direction is found by multiplying by the sine or cosine
of the same angle.

The reason for the importance of this case is easily seen.
Suppose O (Fig. 73) to be a curtain ring sliding on a smooth rod
Ox, and pulled obliquely by a cord with a force R along OR. The
ring can only slide along the rod. In finding whether it will
remain at rest or begin to move we are not helped by resolving R
into components P and Q as in Fig. 72, for then besides the
component P along the rod, the oblique force Q will still have to
be reckoned with. But if R be replaced by a component along the
rod and another at right angles to it, the latter may be left out of
account, since it can produce no effect in the direction of the rod.
It is a great simplification to have thus got rid of all oblique
forces.

163.   III.   Three or more forces.

(a)   Geometrical Method.

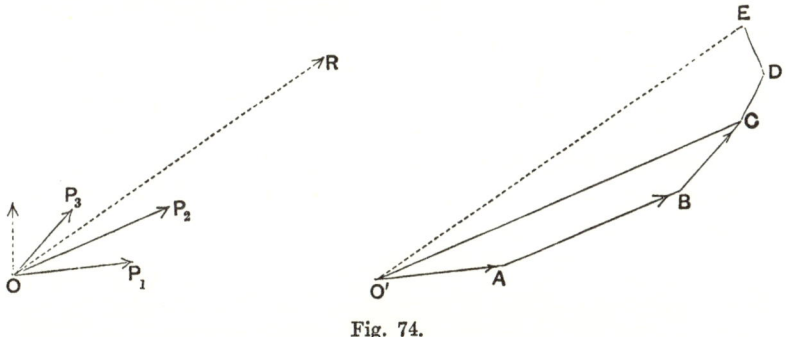

Fig. 74.

Let the forces P_1, P_2, P_3, ... act at O as in Fig. 74.

Make a construction diagram. From any point O' draw $O'A$
to represent P_1; from A draw AB to represent P_2; and so on.
Let DE represent the last force.

Then $O'B$ represents the resultant of P_1 and P_2; $O'C$ the
resultant of $O'B$ and P_3, *i.e.* of P_1, P_2 and P_3; and finally $O'E$ the
resultant of all the forces.

Draw OR equal and parallel to $O'E$. This is the resultant.

164. (*b*) Trigonometrical Method.

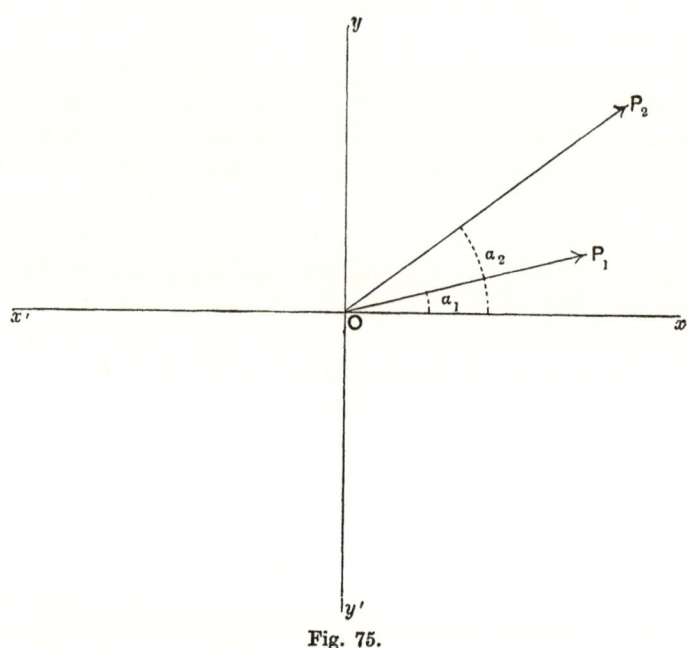

Fig. 75.

Let the forces P_1, P_2, ... act at O.

Choose any direction xOx', and yOy' at right angles to it. Let the forces make angles α_1, α_2, ... with Ox.

Resolve each of the forces P_1, P_2, ... into its components along Ox, Oy.

The components of P_1 are $P_1 \cos \alpha_1$ along Ox and $P_1 \sin \alpha_1$ along Oy,

„   „   „ P_2 „ $P_2 \cos \alpha_2$ „ Ox „ $P_2 \sin \alpha_2$ „ Oy,

and so on for all the forces.

The oblique forces are thus got rid of, and we have only a set of forces $P_1 \cos \alpha_1$, $P_2 \cos \alpha_2$, &c. acting in the same direction along Ox; and another set $P_1 \sin \alpha_1$, $P_2 \sin \alpha_2$, &c. along Oy.

Let the sum of the forces along Ox be X; that of the forces along Oy be Y; so that

$$X = P_1 \cos \alpha_1 + P_2 \cos \alpha_2 + ...,$$
$$Y = P_1 \sin \alpha_1 + P_2 \sin \alpha_2 + ...$$

Then the original set of forces is reduced to two forces X and Y acting at right angles.

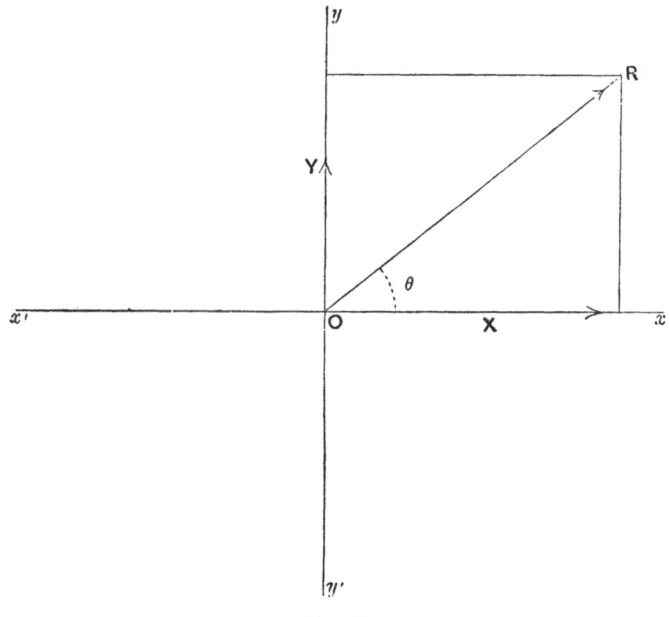

Fig. 76.

The resultant of these is R, where

$$R^2 = X^2 + Y^2 \dots\dots\dots\dots\dots\dots\dots(1),$$

and it makes an angle θ with Ox, such that

$$\tan \theta = \frac{Y}{X} \dots\dots\dots\dots\dots\dots\dots\dots(2).$$

The advantage of this method is that however many forces there may be, X and Y can be written down at once, as the values of the cosines and sines are found from the tables. Then R and θ are easily found from (1) and (2).

165. I. Two Forces.

Two forces acting at a point can only balance, *i.e.* fail to have
The Conditions of a resultant, when they are equal in magnitude and
Equilibrium. directly opposed to each other.

(a)   Graphically.

For it is only when the above conditions are fulfilled that the diagonal of the parallelogram vanishes.

(b)   Analytically.

The Trigonometrical formula for the resultant leads to the same result, as follows:

$$R^2 = P^2 + Q^2 + 2PQ \cos \alpha$$
$$= P^2 - 2PQ + Q^2 + 2PQ(1 + \cos \alpha)$$
$$= (P - Q)^2 + 4PQ \cos^2 \frac{\alpha}{2}.$$

This can only vanish if $P - Q = 0$, and $\cos \frac{\alpha}{2} = 0$, *i.e.* if $P = Q$, and $\alpha = 180°$.

166.   II.   Three Forces.

(a)   Graphically.

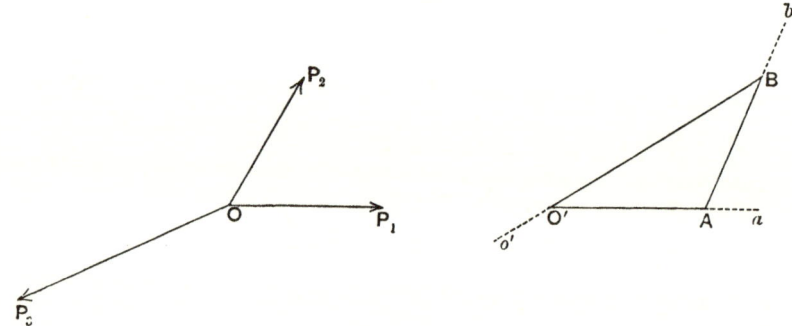

Fig. 77.

Make the force diagram $O'AB$ for P_1, P_2; their resultant is $O'B$. In order that this may balance P_3, P_3 must be represented in magnitude and direction by BO' (taken in the sense of the arrow).

Or, make the force diagram $O'ABC$ for all three forces. Then unless C falls upon O', they will have a resultant $O'C$.

Hence : *In order that three forces acting at a point may be in equilibrium they must be represented in magnitude and direction by the three sides of a triangle taken in order.*

Conversely: *If three forces represented in magnitude and*

direction by the three sides of a triangle taken in order be applied at a point, they will be in equilibrium.

This proposition is known as the Triangle of Forces.

167. (*b*) Analytically.

Father Lami in his *Mécanique* (published in 1687, the year of Newton's *Principia*) gave the Triangle of Forces a Trigonometrical form.   Produce $O'A$, AB, BO' (Fig. 77).

Then $\angle P_1 O P_2 = \angle aAB = 180° - A.$

Similarly, $\angle P_2 O P_3 = \angle bBO' = 180° - B,$

$\angle P_3 O P_1 = \angle o'O'A = 180° - O'.$

For equilibrium

$$\frac{P_1}{O'A} = \frac{P_2}{AB} = \frac{P_3}{BO'}.$$

$$\therefore \frac{P_1}{\sin B} = \frac{P_2}{\sin O} = \frac{P_3}{\sin A},$$

$$\therefore \frac{P_1}{\sin P_2 O P_3} = \frac{P_2}{\sin P_3 O P_1} = \frac{P_3}{\sin P_1 O P_2}.$$

Hence Lami's Theorem : *If three forces acting at a point are in equilibrium, each is proportional to the sine of the angle between the other two.*

168.   III.   Any number of Forces.

(*a*)   Graphically.

Make a construction diagram.   Then there will be a resultant unless the last point returns to the first, and the diagram forms a closed polygon.

Hence : *If any number of forces acting at a point are represented in magnitude and direction by the sides of a closed polygon taken in order, they will be in equilibrium.*

This is known as the Polygon of Forces.

169. (*b*) Analytically.

By the method of § 164 the resultant R is given by

$$R^2 = X^2 + Y^2,$$

where $\qquad\qquad X = P_1 \cos \alpha_1 + \ldots\ldots,$

$$Y = P_1 \sin \alpha_1 + \ldots\ldots.$$

For equilibrium $\qquad\qquad R = 0,$

$$\therefore\; X^2 + Y^2 = 0.$$

But since the squares are necessarily positive, this can only be the case when X and Y are separately zero.

$$\therefore\; X = 0$$
and $\qquad\qquad\qquad\qquad Y = 0$,

i.e. The sums of the resolved parts of the forces in any two directions at right angles must be separately zero.

We are at liberty to choose any two directions at our convenience, for in finding the resultant (§ 164) the directions Ox, Oy were taken arbitrarily.

Both X and Y must be zero. If $X = 0$, there can be no resultant tending to cause motion along Ox. But there may still be an unbalanced force along Oy, and yet no effect produced along Ox, at right angles to it. It is necessary therefore to have $Y = 0$ as well.

The two conditions secure that there shall be no disturbance in either of two mutually perpendicular directions. There cannot then be an unbalanced force in any other direction, since had such an oblique force existed, it must have had components along both Ox and Oy.

Note that (*b*) can be at once deduced from (*a*) by projecting the construction diagram on to any straight line in the plane. For the sum of the projections of the sides of a closed polygon on any straight line is zero.

Conversely, if we project on each of two straight lines, and find the sum of the projections in each case zero, the polygon must be closed.

EXAMPLES.

1. Shew, by a drawing, that if the angle at which two forces are inclined to each other be increased their resultant is diminished.

2. Hence shew that if a picture is hung from a nail by a string fastened to two rings in the top of the frame, the shorter the string the stronger it ought to be. Could the string be stretched so tightly between the rings as to remain straight when placed over the nail?

3. Two forces acting at right angles to each other have a resultant which is double the smaller force. Find its direction.

4. $ABCD$ is a parallelogram, and AB is bisected in E; prove that the resultant of the forces AD, AC is double the resultant of AE, AC.

5. $ABCD$ is a quadrilateral, and E the point of intersection of the lines joining the middle points of opposite sides; O is any point. Prove that the resultant of forces OA, OB, OC, OD is equal to $4OE$.

6. Two forces act at a point. Shew that if, when one of the forces is reversed, the resultant is at right angles to the direction of the resultant before the change, the forces are equal.

7. $ABCD$ is a quadrilateral. Shew that if four forces represented by AB, AD, CB, CD be applied at a point, their resultant will be represented by four times the line joining the middle points of the diagonals.

8. Find the magnitude and direction of the resultants of the following pairs of forces (in pound weights):

      (1)   24 lbs. and 7 lbs. acting at right angles,

      (2)   7 lbs. and 8 lbs. at an angle of 60°,

      (3)   11 lbs. and 14 lbs. at 120°,

      (4)   6 lbs. and 8 lbs. at 52°.

9. Forces 7, 12, 3, 11 act at a point, the first due East; the second North-East, the third North; and the fourth 60° west of North. Find their resultant.

10. Forces 1, 2, 3, 4, 5, 6, 7, 8 act at a point, the angle between each force and the next being 47°. Find the magnitude of the resultant, correct to two places of decimals, and its direction.

11. Find the resolved part of a force of 60 lbs. in a direction inclined 40° to the force.

12. A canal-boat is pulled by a rope 60 feet long, and the boat is 30 feet from the towing path. If the horse pulls with a force of 120 lbs. weight, what is the force urging the boat forward ?

13. A captive balloon capable of raising a weight of 400 lbs. is anchored at a height of 400 feet by a rope 500 feet long. Find the strain on the rope and the horizontal pressure of the wind on the balloon.

14. A 50 lb. weight hangs by a wire 13 feet long. What horizontal force is required to draw it aside 5 feet from the vertical through the point of suspension, and what will then be the tension of the wire ?

15. A body of weight 15 lbs. is placed on an inclined plane 3 feet high and 5 feet long. Find the components of its weight along and perpendicular to the plane.

16. Explain how a boat can sail almost in the eye of the wind, by setting the sail between the direction of the wind and the boat's course.

(The velocity of the wind may be resolved into a component parallel to the sail, which has no effect, and a component perpendicular to the sail, which exerts a pressure on it. This pressure may again be resolved into components parallel and perpendicular to the boat's length. The former is the propelling force ; the latter causes leeway, which is made as small as possible by using a keel or centreboard to resist sideway motion. Draw a diagram shewing the two resolutions, and shew that if P be the pressure which the wind would exert on the sail if at right angles to it, a, β the inclinations of wind and sail to the keel of the boat, then

$$\text{headway force} = P \sin (\beta - a) \sin \beta,$$
$$\text{leeway force} \quad = P \sin (\beta - a) \cos \beta.)$$

17. Explain how a kite is sustained in air, and shew by a drawing that the perpendicular to the kite must lie between the direction of the string and the vertical.

18. Forces 24, 7, and 25 lbs. weight balance at a point. Shew that two of them are at right angles.

19. A 50 lb. weight is hung from two points by strings inclined 30° and 45° to the vertical. Find the tensions of the strings.

20. A picture weighing 8 lbs. is hung by a string passing over a nail and attached to two rings in the top of the frame. Find the tension of the string when the two portions are inclined at an angle of (1) 60°, (2) 120°.

21. Find the ratio of the "power" to the "weight" in the inclined plane by resolving along and at right angles to the plane. Find also the pressure on the plane.

(This is the best way of treating the inclined plane, and serves equally well when the force is not parallel to the plane.)

22. A weight W rests on an inclined plane, inclination a. Find the force required to sustain it, and the pressure on the plane,

(1) when the force acts horizontally ;

(2) when its direction makes an angle ϵ with the plane and above it.

23. Shew that if the strings supporting a single moveable pulley are inclined at θ to the vertical, $P = W/2 \cos \theta$.

24. Weights P and Q rest on the upper edge of a smooth vertical circle, and are connected by a string, running round the edge, whose length is a quadrant of the circle. Find the position of equilibrium, and the tension of the string.

(Write down the conditions of equilibrium first for P and then for Q, by resolving along the tangent and at right angles to it in each case, assuming that the radius to P makes an angle θ with the horizontal. The resulting equations determine θ and T.)

25. Two men raise a cask weighing 300 lbs. from a cellar to the street by drawing it up planks inclined 30° to the horizon by means of two ropes fastened to the wheels of a dray in the street, passed down the planks, under and round the barrel, and pulled parallel to the planks. What is the least force each man must exert ?

26. A conical pendulum consists of a ball weighing 5 lbs. suspended by a string 4 feet long. If the ball is projected so as to describe a horizontal circle twice in three seconds, what will be the inclination of the string to the vertical, and what will be its tension ?

In what time must the ball revolve in order that the string may be inclined 30° to the vertical?

(The acceleration of the ball to the centre may be calculated by § 77. The resultant force on the ball, *in poundals*, in order that it may go on describing the circle is the product of this acceleration into the mass of the ball. We may then either :

(1) express the condition that this force is the horizontal component of the tension of the string, while its vertical component is equal to the weight ;

or (2) observing that if this force were reversed in direction, it would be in equilibrium with the tension and the weight, treat the problem as if the ball were at rest under the weight, the tension, and the reversed resultant force.)

27. A conical pendulum with a string of length l makes n revolutions per second. Shew that the inclination of the string is a, where $\cos a = \dfrac{g}{4\pi^2 n^2 l}$.

28. Prove that in the conical pendulum the time of revolution is $2\pi \sqrt{\dfrac{h}{g}}$, where h is the vertical depth of the revolving ball below the point of support.

29. Apply this result to the governor of a steam engine, and shew that for an engine making 60 revolutions per minute the depth of the balls below the point of support must be about 9·78 inches.

30. Why is the outside rail of a railway track raised above the inside rail at a curve?

Shew that if a train runs round a curve of radius r feet with velocity v, the floor of the carriage should be inclined at an angle whose tangent is v^2/gr in order that there may be no lateral thrust on the rails.

31. Shew that on a 5-foot track, round a curve of one-eighth of a mile radius, for a mean velocity of 30 miles an hour, the outside rail ought to be raised between 5 and 6 inches above the level of the inner rail.

CHAPTER XVIII.

FORCES ACTING ANYWHERE IN A PLANE.

170. I. RESULTANT of Two Forces acting at different points.

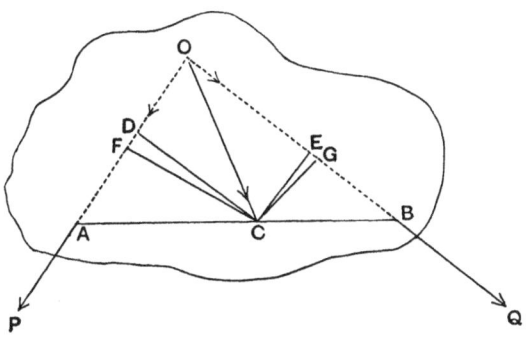

Fig. 78.

Let the forces P and Q act on a body at A and B.

To fix the ideas, suppose that the body is a flat board lying on a horizontal table, and that the forces are applied by cords, attached to pins at A and B, and carrying weights at their free ends, which hang over the edge of the table.

Let the directions of P and Q, produced if necessary, meet in O. The effect of P and Q will not be altered if they are applied at O in the same directions, instead of at A and B.

This principle, that the effect of a force is the same, at whatever point in its line of action it is applied, is known as the principle of the *Transmissibility of Force.* It may be regarded as an axiom directly based on experience. In fact we feel that if the cords by which P and Q are applied are prolonged to O and fastened to a pin there, the pins at A and B may be taken out without disturbance.

(Or the principle may be deduced from some other axiom of experience, such as the Third Law of Motion. All actual bodies undergo slight changes of shape on the application of force. The idea of a *perfectly rigid body* is a mathematical fiction, useful because most of the solids known to us approximate so closely to it that in Statics, where we are concerned with the external relations between different bodies, we can greatly simplify our theorems if we are content to ignore the very small internal displacements and the corresponding (often great) internal forces that are called into play. When these are taken account of, we enter on the Theory of Elasticity.

Consider, as the simplest case, a fine rubber thread kept stretched by two forces applied at the ends. Every particle of the thread is drawn apart from those on each side of it till the forces of cohesion so developed are sufficient to prevent further stretching. By the Third Law the forces between each pair of particles are then equal and opposite. The stretching force applied to one of the end particles must, for equilibrium, exactly balance the internal pull of the second particle upon the first; and so on throughout the string, till the internal pull upon the last particle balances the external force applied to the other end. The pull is thus transmitted by the *stretched* thread so as to balance the exactly equal pull at the other end.

In the case of the elastic thread the displacements would be so large that they must be taken account of; but the internal forces, though everywhere equal to the external pulls, may be ignored, since they occur in equal and opposite pairs. The same process goes on in the solids contemplated in Statics as *rigid bodies*. But the displacements are so small as not to affect the configuration, and so may be left out of account. The internal forces may be ignored for the same reason as before; and this is true even when the line of action of the transmitted force passes outside the body. For let two equal forces act at A and O (Fig. 79) in opposite directions. Then the internal reactions between the parts of the body, which hold it together, are, by the Third Law, at every point, whether in the line AO or elsewhere, equal and opposite. Therefore the whole set, including the two equal forces at A and O, will balance, just as they would if both the forces were applied at the point O. Hence the effect of a force at A is the same as if it were applied at O, another point in its line of action.)

The resultant of P and Q acting at O passes through O. Let it cut the line joining AB in C, and take OC, which already represents it in direction and point of application, to represent it also in magnitude.

Draw CD, CE parallels to OB, OA; then OD, OE will represent the components P and Q on the scale of the resultant OC.

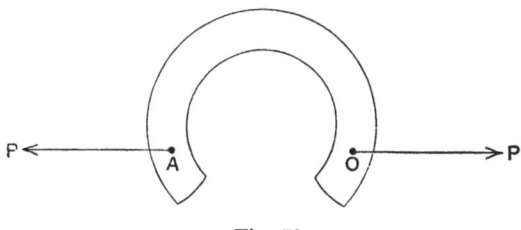

Fig. 79.

Drop CF, CG perpendiculars to OA, OB. Then, area of the parallelogram

$$CDOE = OD \times CF = OE \times CG.$$

$$\therefore \; P \times CF = Q \times CG.$$

Thus the resultant cuts AB in a point C such that *the moments of P and Q about C are equal and opposite.*

171. Since we might have taken any length OC to represent the resultant, this property must hold for all points on the resultant; or, what comes to the same thing:

The algebraical sum of the moments of two concurrent forces about any point on the line of action of their resultant is zero.

172. The property just proved is a particular case of a

Varignon's Theorem of Moments.

theorem communicated to the Paris Academy by Varignon in 1687 (the year of the *Principia* and of Lami's theorem).

The moment of the Resultant of two co-planar Forces about any point in their plane is equal to the (algebraical) sum of the moments of the Forces.

Consider two forces AB, AC and their resultant AD.

The moment of the force AB about the point O is the product of AB by the perpendicular from O on AB, *i.e.* twice the area of the triangle OAB. Similarly the moments of AC, AD will be twice the triangles OAC, OAD.

We have to shew that

$$OAD = OAB + OAC$$

when O is outside the angle between the forces (Fig. 80 a),

and $$OAD = OAB - OAC$$

when O is inside (Fig. 80 b).

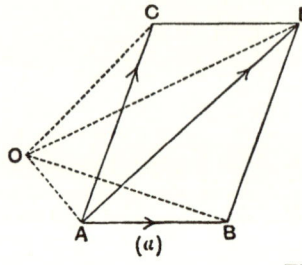

Fig. 80.

The perpendiculars from the vertices B, C, D of the three triangles OAB, OAC, OAD upon their common base OA are equal to the projections of AB, AC, AD on a line at right angles to OA.

But the projection of AD on any line is equal to the algebraic sum of the projections of AB and BD; or of AB and AC which is equal and parallel to BD.

∴ the area OAD is equal to the algebraic sum of the areas OAB, OAC.

The + sign is obviously to be taken in Fig. 80 (a), and the − in Fig. 80 (b).

If O is on the line of the resultant, the moment of the resultant and therefore the sum of the moments of the forces, is zero.

Note. The Moment of a force about a point may be conveniently represented by the area (or double the area) of the triangle formed by joining the point to the ends of the line representing the force.

173. If the forces P and Q (§ 170) are parallel, their directions will not meet, and our construction fails.

Parallel Forces.

This is a case for employing the principle of Continuity. Let us start with the figure of § 170, and gradually bring the forces to parallelism, making them both approach the direction perpendicular to AB. The parallelogram $CEOD$ becomes more and more lozenge-shaped (Fig. 81), and the diagonal is more and more nearly equal to the sum of the sides. The law of moments,

$$P \times CF = Q \times CG,$$

remains true, but CF and CG approach the values CA, CB.

We can see what will happen in the limit, when the forces become really parallel.

(1) The Resultant becomes the sum of the forces, so that

$$R = P + Q.$$

(2) It is parallel to the forces.

(3) It cuts AB in C, so that

$$P \times AC = Q \times BC.$$

It is thus completely determined.

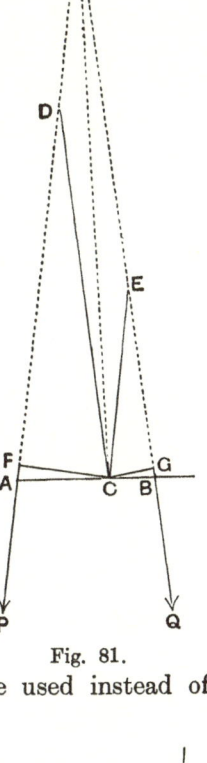

Fig. 81.

When the forces are parallel, the segments of any oblique line $A'CB'$ through C may be used instead of ACB, since

$$\frac{A'C}{B'C} = \frac{AC}{BC} = \frac{Q}{P},$$

$$\therefore P \times A'C = Q \times B'C.$$

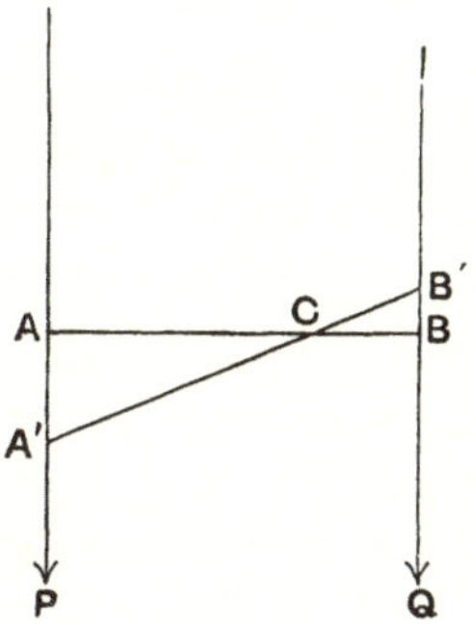

174. One of the most familiar cases

The Principle of the Lever.

is when the two forces act on a bar or other body which is only free to turn on a pivot. To find the relation between the forces and their distances so that they should have

Fig. 82.

equal power to turn the bar about the pivot was Archimedes' famous problem of the Lever. (§ 3.)

The difficulty of judging between unequal forces at unequal distances vanishes when only one force is applied to the bar, for it will certainly turn it one way or the other *unless the force goes through the pivot.*

But we can now replace the two forces by their single resultant. Then if C is the pivot the resultant must, in the case of equilibrium, go through C, and therefore, whether the forces intersect or are parallel, if they are to have equal torques, *i.e.* tendencies to turn the bar about the pivot, the product of each into the perpendicular distance from the pivot must be the same. This product must therefore be the proper measure of the torque of a force about a point. As we have seen, Leonardo called it the Moment of the force about the point.

The principle of the Lever thus follows from the Parallelogram of Forces. We might go on to deduce from it all that has been given in §§ 6—33 about the Centre of Gravity, the Balance, Wheel and Axle, and the Pulleys.

175. The student may deduce from Fig. 83 what will happen when the forces

Parallel Forces in Opposite Directions. act in opposite directions. Let Q be the larger; then

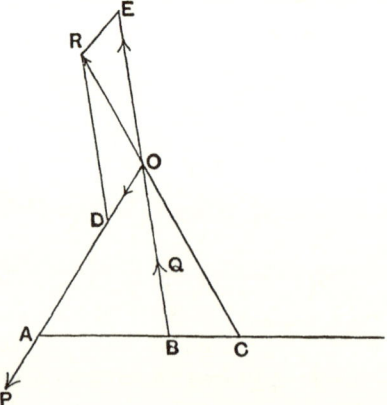

(1) $R = Q - P$.

(2) R is parallel to P and Q and cuts AB outside, beyond the larger force.

(3)   Since C is on the resultant, by Varignon's Theorem, the moments of the forces about C are equal and opposite. If they are perpendicular to AB,

$$P \times AC = Q \times BC,$$

Fig. 83.

and this is extended as before to an oblique line $A'CB'$.

(Or the case of unlike forces may be deduced from that of like forces as in § 13.)

176. An important case of failure of the method for finding
Couples.   the resultant of two forces remains to be
considered.   If the forces are unlike in direction
and *equal*, the resultant $Q - P$ vanishes.   Moreover no point
C can be found, outside AB, which will make $P \times AC = P \times BC$.
When Q is very nearly equal to P, C has to be a long way off.
For equality C would have to be at an infinite distance.

In fact a pair of equal parallel forces acting in opposite directions has no single resultant, and cannot be balanced by any single force.   They have no tendency to move a body from one place to another, which could be met by a single force; but they tend to turn it round in its place; to give it a twist.   Such are the forces applied by the thumb and finger to the wings of a screw nut; or by the hands to the bar of a copying-press; or to a capstan by two men working on opposite sides of it.

A pair of equal, unlike, parallel forces is called a *Couple*.   The perpendicular distance between the forces is called the *Arm* of the Couple.

A Couple has no single resultant, and no single force can balance it.   But it has a twisting tendency, or *Torque*, measured by its moment.

177. The theory of Couples was introduced by Poinsot, and
The Properties   affords a beautiful method of simplifying compli-
of Couples.   cated systems of forces.

(In thinking about couples it should be borne in mind that they are here supposed to be applied to a rigid body kept at rest by certain forces, and that the couples considered form part of the system of forces maintaining the equilibrium.   It is shewn in works on Rigid Dynamics that the effect of a couple applied to a rigid body otherwise free to move is to set it rotating about an axis passing through its centre of gravity, but not necessarily perpendicular to the plane of the couple.)

Let the forces in Fig. 84 be not quite equal; suppose the force at A to be the weight of one pound ($= 7000$ grains), and

12—2

that at B to be one pound and one grain. Let AB be one foot. Then their resultant will be a force of one grain, directed upwards and applied at a point O, 7000 feet away to the right of B.

If these forces are applied to a rod pivoted at O, there is equilibrium. But if the pivot be anywhere else, there will be a tendency to turn measured by the moment of the resultant about the pivot. For instance, if we take for unit of moment the moment of a force of one pound about a point distant one foot from its line of action, and if the pivot is 20 feet to the right of B, the moment will be $6980 \times \frac{1}{7000} = \frac{6980}{7000}$ units. The moment about a pivot 10 feet to the right of B is $\frac{6990}{7000}$; about B it is $\frac{7000}{7000} = 1$; about A, $\frac{7001}{7000}$.

It is clear that for pivots anywhere in the neighbourhood of the forces the moments are all very nearly equal—equal, indeed, to the product of the force at B (one pound) by the distance AB (one foot). To make a difference of so much as one per cent., the pivot must be at least 70 feet away from B. In spite of the smallness of the resultant its moment remains considerable owing to the distance at which it acts. But then a considerable change in that distance is required to produce any marked alteration in the value of the moment.

Proceeding to the limit when the forces are exactly equal (say one pound each), we see that (1) the resultant utterly vanishes ; (2) the moment remains finite (equal to one unit); (3) the moment is the same wherever the pivot is placed in the plane of the couple.

This might have been deduced directly from a consideration of the forces.

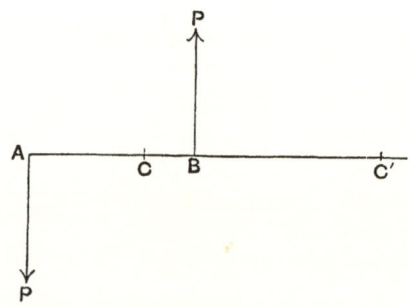

Fig. 84.

For draw *any* line AB cutting the forces at right angles, and take moments about a point C in AB. Then if C is between A and B,

Moment of Couple $= P \times AC + P \times BC = P \times AB$;

and if C is outside, as at C',

Moment of Couple $= P \times AC' - P \times BC' = P \times AB$.

Thus $P \times AB$ is the moment about any point in AB; and AB may be drawn anywhere.

Hence, (1) *the Moment of the Couple is the same about every point in its plane; and is measured by the product of either of the forces into the perpendicular distance between them, i.e. into the arm.*

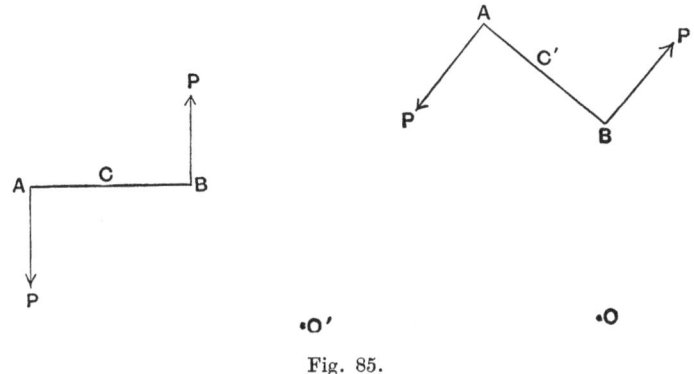

Fig. 85.

Consider two couples with equal forces and arms, but in different positions with regard to a point O. Neither of them has any resultant. The only effect of each is a torque, or tendency to turn the body about O; and this is measured by the product of the force and arm, which is the same for each. The couples are therefore equivalent.

Or look at it in this way. It is easy to find a point O' which is placed with regard to the couple C precisely as O is with regard to C'. The effect of C' about O is then the same as that of C about O'; or by (1), of C about O.

Hence, (2) *a couple may be turned through any angle without altering its effect,* and

(3) *a couple may be removed to any other position in the plane without altering its effect.*

(4) *Since the whole effect of a couple is measured by its moment, a couple may be replaced by any other couple in the same plane having an equal moment.*

(Or, directly from the forces, supposed parallel.

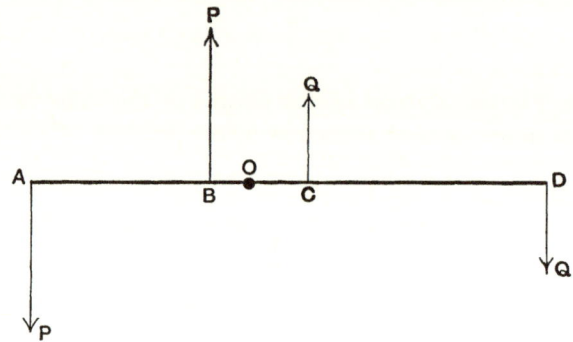

Fig. 86.

Let PP, QQ be two couples of equal moment, so that
$$P \times AB = Q \times CD,$$
but acting in opposite senses.

The resultant of P at B and Q at C is $P + Q$ acting at a point O such that
$$P \times BO = Q \times OC.$$
But $\qquad\qquad P \times AB = Q \times CD,$

$$\therefore P \times (AB + BO) = Q \times (OC + CD)$$
or $\qquad\qquad P \times AO = Q \times OD.$

$\therefore$ the resultant of P at A and Q at D is $P + Q$ acting at the same point O, but in the opposite direction.

The couples therefore balance, so that a couple $P \times AB$ is equivalent to another couple $Q \times CD$ of equal moment acting in the same sense.)

(5) *Any number of couples in a plane may be replaced by a single couple of the same total moment.*

Since the only effect of a couple is its torque, measured by its moment, this follows from the physical independence of forces implied in the Second Law of Motion. But for the sake of the importance of the subject we will now deduce this (and incidentally all the previous propositions) directly from the parallelogram of forces. The student may draw the figure for himself.

178. Consider two couples with forces PP' and QQ', and in order to take the most general case let the forces of the one couple be not parallel to the forces of the other couple.

If the lines of action of the four forces be produced, they will form a parallelogram $ABCD$, the sides AB and CD being the lines of action of the forces P and P' of the one couple, and the sides AD and CB being the lines of action of the forces Q and Q' of the other couple.

But the resultant of P and Q is a force R acting through A, and having a moment about any point which is the algebraical sum of the moments of P and Q.

Also the resultant of P' and Q' is a force R' acting at C, and equal and opposite to R.

Thus the two forces R and R' form a couple which is in all cases the resultant of the two original couples. And since the moment of a couple is the sum of the moments of its component forces, we see at once from Varignon's Theorem that the resultant couple has a moment which is the algebraical sum of the moments of the original couples.

Again, when the moments of the original couples are equal and opposite, the resultant couple has no moment, that is, the original couples balance each other. In this case R and R' are not only equal and opposite, but they act along the same line.

179. II. Three or more Forces.

Any number of forces acting at different points in a plane can always be reduced to a single force acting at any point and a couple.

Let P be one of the forces, acting at a point A. Take any convenient origin O, and apply at it two equal opposite forces each equal and parallel to P. This balancing pair of forces will not alter the effect.

But the three forces are now equivalent to

   (1)   a force P acting at O, and

   (2)   a couple of moment $P.p$.

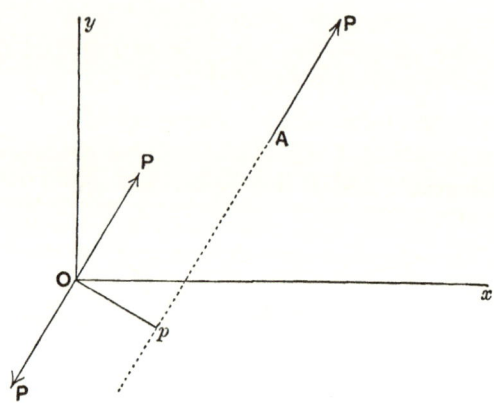

Fig. 87.

Let the same be done for all the other forces. The system reduces to

   (1) the forces transferred from their actual points of application, and all acting at O;

   (2) a set of couples.

The resultant of (1) may be found by § 164. The couples (2) are equivalent to a single couple of moment equal to the sum of their moments.

We may carry the simplification one step further, and reduce the system to a single force, *or* a couple.

For if the resultant force does not vanish, let the couple be replaced by another couple of equal moment, but with forces each equal to the resultant force. Let this couple be turned round its axis till one of its forces is directly opposed to the resultant force. Then these two vanish, and the system reduces to the other force of the couple; *i.e.* to a single force.

In the case where the resultant force vanishes the system obviously reduces to the couple.

180. It is often more convenient to choose two axes of reference, and to resolve each force into its components in these directions before transferring to the origin.   Thus:

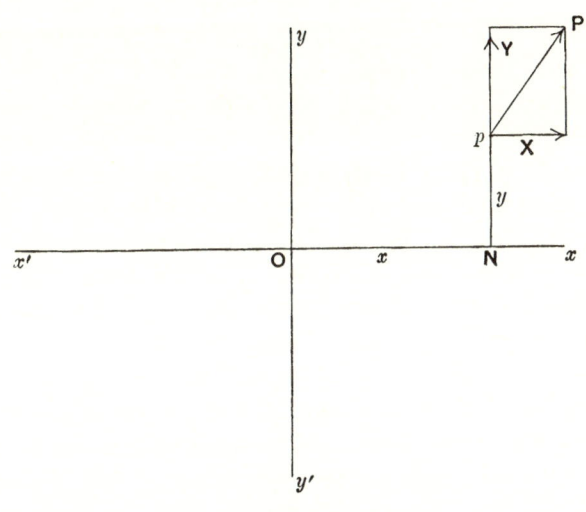

Fig. 88.

Let the position of p, the point of application of the force P, be fixed by the co-ordinates $ON = x$, $Np = y$; and let the components of P parallel to the axes be X and Y.

Then X may be replaced by

(1) X acting at O along Ox, and

(2)   a couple $-Xy$.

Y may be replaced by

(1) Y acting at O along Oy, and

(2)   a couple $+Yx$.

Let the same be done for all the forces, and let us indicate the sum of all products like Xy by ΣXy.

The system reduces to ΣX and ΣY acting at O; and a set of couples $\Sigma (Yx - Xy)$.

The resultant of ΣX and ΣY will be a force R, and a single couple G may be found whose moment is equal to $\Sigma (Yx - Xy)$.

181.   I.   Two Forces.

Conditions of Equili-
brium for any Forces
in one Plane. These must act along the same line in opposite directions, and be equal in magnitude.

182.   II.   Three Forces.

If three forces are in equilibrium, they either meet in a point, or are parallel.

For if two of them meet, their resultant goes through their meeting point, and can only be balanced by a force passing through that point; but if two of them are parallel, their resultant is parallel to them, and therefore so is the force which is to balance them.

Very many problems occur in which the lines of action of two out of the three forces are known, and the solution depends on making the third force pass through their point of intersection. The magnitudes of the forces are then generally given by Lami's Theorem, or some geometrical application of the Triangle of Forces.   In the case of parallel forces we have

$$R = P + Q$$

and $$P \times AC = Q \times BC.$$

183.   III.   Any number of Forces.

Let them be reduced to a single force R acting at any point; and a couple G.

Since a Couple cannot be balanced by a force, it is necessary for equilibrium that both the force and the couple should vanish separately.

Thus

$$\left. \begin{aligned} R &= 0 \\ G &= 0 \end{aligned} \right\}.$$

These are equivalent to

$$\left. \begin{aligned} X &= 0 \\ Y &= 0 \end{aligned} \right\} \ (\S\,169)$$

and $$\Sigma\,(Yx - Xy) = 0,$$

i.e. (1) *the sums of the resolved parts of all the forces in any two directions at right angles must be separately zero; and*

(2) *the sum of the moments of all the forces (or of their components) about any point must vanish.*

The conditions (1) secure that there shall be no movement
of translation, while (2) must be satisfied if there is to be no
rotation.

184. The general procedure in solving a Statical problem
Method of Solving   is as follows:
Statical Problems.   (1) Draw a figure and see that all the external
forces, as well as any internal reactions that are to be considered,
are properly represented.

(2) Select some body or system of bodies, and write down the
conditions of equilibrium for it; *i.e.* choose any convenient axes
at right angles; resolve the forces acting on the system along
them, and equate the sums of the resolved parts to zero; and
"take moments" about any convenient point, equating the sum
of the moments to zero.

It is most important, as in Dynamics (§ 128), to settle quite
definitely what is to be the system considered when writing down
each equation.

The two equations of the resolved parts, and the one of
moments, in general suffice to determine two co-ordinates of some
point in each body, and one angle fixing its azimuth about the
point. For each unknown reaction, such as a pressure between
bodies in contact, or the tension of a string, a geometrical relation
can be written down; so that there will be as many equations as
quantities to be found.

By choosing axes at right angles to some of the forces, and by
taking moments about points through which their directions pass,
we can often prevent forces whose values are unknown, or not
wanted, from entering into the equations, and obtain what we
want from *one* of the equations of resolution, or from the equation
of moments *alone*. A proper choice of axes and points for taking
moments will greatly simplify most problems.

EXAMPLES.

1. A uniform beam of weight W can turn about a hinge at one end A, and is drawn aside from the vertical by a horizontal force P applied to the other end B. Find the position of equilibrium and the reaction at the hinge.

Consider the beam. Three forces act on it, P, W, and the reaction R at the hinge. Hence R must go through the meeting point of P and W.

Let $2a$ be the length of the beam ; θ, ϕ the inclinations to the vertical of the beam and of the reaction R.

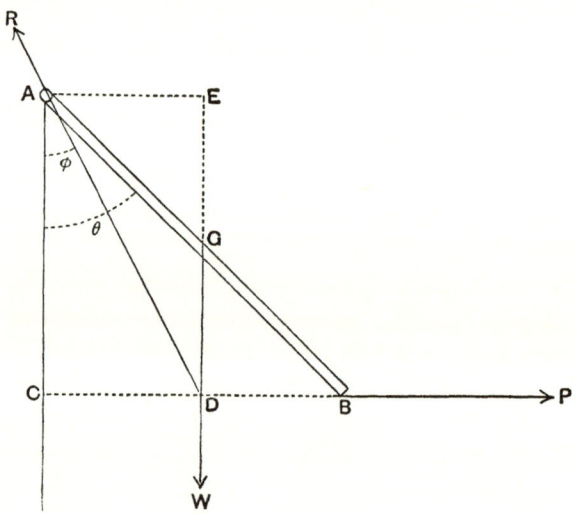

(1) By Lami's Theorem

$$\frac{P}{\sin \phi} = \frac{W}{\cos \phi} = \frac{R}{\sin 90°}.$$

$$\therefore \ \tan \phi = \frac{P}{W}.$$

Whence R is known.

By Geometry

$$\tan \phi = \frac{CD}{AC} = \frac{AE}{AC} = \frac{a \sin \theta}{2a \cos \theta} = \tfrac{1}{2} \tan \theta.$$

$$\therefore \ \tan \theta = 2 \tan \phi = \frac{2P}{W}.$$

(2)   By the general conditions of equilibrium.

Without necessarily assuming that R goes through D, we have, by resolving horizontally and vertically,

$$R \sin \phi = P,$$

$$R \cos \phi = W,$$

whence R and ϕ are known.

The inclination of the beam is found directly by taking moments about A, when R does not come in.   Thus

$$P \cdot 2a \cos \theta = W \cdot a \sin \theta$$

and

$$\tan \theta = \frac{2P}{W}.$$

2.   A beam of length $2a$ and weight W turns on a hinge at one end A, and is supported at an inclination θ to the vertical by a string, attached to its other end B.   The string passes over a smooth pulley at C, and sustains a weight P. C is vertically above the hinge A, and distant b from it.   Shew that

$$\cos \theta = \frac{(4P^2 - W^2)b^2 - 4W^2 a^2}{4ab\,W^2}.$$

3.   One end A of a beam AB, weight W, rests against a smooth vertical wall, and a cord attached to the other end B passes over a smooth pulley at C, vertically above A, and sustains a weight P.   Find the pressure on the wall.

4.   A ladder 30 feet long and weighing 56 lbs. rests, at an inclination of $60°$ to the horizon, with its lower end against the foot of a wall.   It is sustained by a rope attached to a rung 20 feet from the bottom, and carried horizontally to a staple in the wall.   The centre of gravity of the ladder is 12 feet from the bottom.   Find the tension of the rope, and the direction of the reaction at the bottom of the ladder.

What would the tension become if a man weighing 160 lbs. climbed out to the top of the ladder ?

5.   A beam is placed across a smooth horizontal rail, and rests with its lower end against a smooth vertical wall, the distance of which from the rail is one-sixteenth of the length of the rod ; shew that the rod must be inclined $60°$ to the horizon.

6.   A plank 12 feet long and weighing 30 lbs. stands on the ground, leaning against a smooth horizontal rail at an inclination of $60°$ to the ground, with 3 feet of its length projecting beyond the rail.   Find the direction and magnitude of the reaction on the ground.

7. A rod whose centre of gravity divides it into two parts a and b is placed inside a smooth sphere of such a radius that the rod subtends an angle $2a$ at the centre. Find the inclination θ of the rod to the horizon.

(This is a type of [many problems in which beams or other heavy bodies are supported by two forces. We have done with the mechanics of the question the moment we see that the c.g. must be vertically under the meeting point of the forces,—in this case the centre of the sphere.

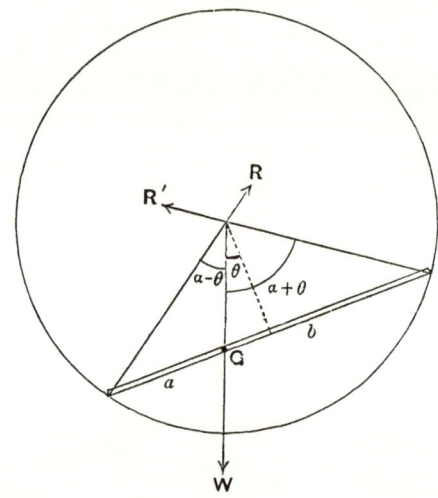

If we drop perpendiculars from the ends of the rod on the vertical through G, the fact that this vertical passes through the centre at once gives the result that

$$\sin (a - \theta) : \sin (a + \theta) = a : b,$$

and finally

$$\tan \theta = \frac{b - a}{b + a} \tan a.$$

Once θ is known, Lami's Theorem gives the pressures.)

8. A beam of weight W whose centre of gravity divides it into portions a and b, rests with its ends on two smooth planes inclined towards each other and making angles a, β with the horizontal. Find the inclination of the beam to the horizon and the pressures on the planes.

(Solve as in Question 7, and also solve by resolving the forces and taking moments.)

9. A beam whose centre of gravity divides it into two portions a and b is suspended by a rope of length l attached to its ends and passed over a smooth peg. Find the inclination of the beam to the vertical and the tension of the rope.

10. A sphere of weight W rests between a vertical plane and a plane inclined at an angle a to the horizon. Find the pressures on the planes.

11. Two spheres, each of one inch radius, rest inside a sphere of 3 inches radius. Shew that the pressure between the small spheres is one-half the pressure of either small sphere on the large one.

12. A sphere is hung from a hook in a vertical wall by a string equal in length to the radius. Find the inclination of the string, its tension, and the pressure on the wall, if the sphere weighs 10 lbs.

13. A square is hung up with its plane perpendicular to a smooth vertical wall by a string attached to one corner and to a hook in the wall, the length of the string being equal to a side of the square. Shew that the distances of the three corners from the wall are as $1 : 3 : 4$.

14. A carriage wheel of radius r and weight W is to be dragged over an obstacle of height h by a horizontal force applied to the centre of the wheel. Shew that the force must be slightly greater than

$$F = W . \frac{\sqrt{2rh - h^2}}{r - h} .$$

15. A step ladder weighing 40 lbs. has two equal legs, each 10 feet long, hinged at the top and joined by a cord 6 feet long at the bottom. It rests on a smooth plane and a weight of 160 lbs. is placed on the top. Find the tension of the cord.

16. One end of a uniform ladder, 84 lbs. weight, rests against a smooth vertical wall at a height of 12 feet from the ground, and the other rests on the ground 10 feet from the wall. Find the pressure on the ground.

17. The two legs of a light step ladder are connected by a smooth joint at the top and a cord at the bottom. The ladder stands on a smooth floor with one leg, which is 3 feet long, vertical. A man of 11 stone weight stands on the other leg at a height of 2 feet above the ground. Find the pressure on the vertical leg. What is the tension of the cord?

18. The sides of a triangular framework are 13, 20, and 21 inches; the longest side rests on a smooth horizontal table, and a weight of 63 lbs. is suspended from the opposite angle. Find the tension in the side on the table.

19. Two small heavy rings of weights W and W', connected by a light string, slide on two wires in the same vertical plane making equal angles a with the horizon. Shew that the inclination θ of the string to the horizon is given by

$$\tan \theta = \frac{W - W'}{W + W'} \cot a.$$

20.   Forces of 3, 5, 7, 9 lbs. weight act along the sides AB, BC, CD, DA of a square, each of the sides being one foot long.   Find their resultant.

(Assume that it cuts the sides AB, AD at points distant x and y from A. Then express the conditions that the moments about these points are zero. This will determine two points on the resultant.   Shew that it is parallel to the diagonal AC.)

21.   Forces of 1, 2, 3, 4, 5, 6 lbs. weight act along the sides AB, BC, &c. of a regular hexagon, taken in order.   Find the single force acting at A and the moment of the couple which together are equivalent to the system.

22.   The resultant of three forces P, Q, R, which act along the sides of a triangle ABC taken in order, passes through the centres of the circumscribed and inscribed circles.   Shew that

$$\frac{P}{\cos B - \cos C} = \frac{Q}{\cos C - \cos A} = \frac{R}{\cos A - \cos B}.$$

23.   Four forces are completely represented by the sides AB, AD, CB, CD of a quadrilateral $ABCD$.   Shew that they are equivalent to a couple consisting of two forces through A and C each equal and parallel to the other diagonal.

24.   Solve Question 20 by finding the single force acting at A, and the couple which together with it is equivalent to the system.   Then change the couple to another of the same moment with forces each equal to the single force ; and turn it round A till one of its forces balances the single force, leaving the other as the resultant.

CHAPTER XIX.

FRICTION.

185. ON the principle of dealing with one difficulty at a time (§ 101) we have hitherto supposed that the surfaces of the bodies in contact were perfectly smooth, so that the only possible reaction between them was a direct thrust perpendicular to the surface of contact. This is never the case in practice. Wherever there is tendency to motion of one body which presses upon another, friction is called into play to oppose it.

186. The effects of friction were first investigated by Coulomb. He used a weighted slider set upon a horizontal surface, with a cord from it carried over a pulley to support a scale-pan.

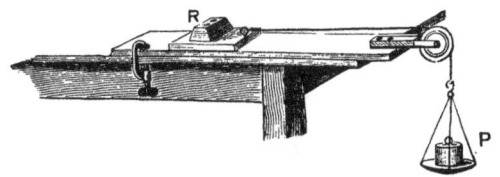

Fig. 89.

When no weight is attached to the string, the upward pressure from the table balances the weight of the slider, and the system remains at rest. If weights be gradually placed in the scale-pan, the slider at last gets into motion, and then proceeds with accelerated velocity.

Thus up to a certain limit just enough friction is called into play to prevent motion. If the weight P exceeds this limit, motion must ensue, and since it is accelerated, the friction during

C.   13

motion is slightly less than it was before starting. Coulomb accordingly made his measurements by first starting the slider, and then finding the weight P required to keep it in uniform motion.

He deduced from his experiments the following laws for the limiting friction:

(1) The friction varies as the normal pressure when the materials of the surfaces in contact remain the same.

(2) The friction is independent of the extent of the surfaces in contact so long as the normal pressure remains the same.

(3) The friction is independent of the velocity when the body is in motion.

These laws are only approximately true, and should not be relied on for values of the normal pressure or of the velocity outside the limits of those employed in the experiments from which the friction was determined. But if we may assume their truth, then the friction is given by the equation $F = \mu R$. μ is called the *coefficient of friction* for the given materials.

The best way to form an idea of the value of μ is to make actual experiments with the simple apparatus of Coulomb. Rankine gives the results of some experiments as follows:

For iron on stone μ varies between ·3 and ·7,

timber on timber	,,	,,	·2	,,	·5,
timber on metals	,,	,,	·2	,,	·6,
metals on metals	,,	,,	·15	,,	·25.

187. Let a body be placed on a horizontal plane, and let the plane be gradually tilted till the body is just on the point of sliding. The friction will act up the plane and be μ times the pressure.

The Inclined Plane affords another means of finding μ.

Resolving along and perpendicular to the plane, we have

$$W \sin \alpha = \mu R,$$

$$W \cos \alpha = R,$$

$$\therefore \quad \mu = \tan \alpha.$$

The angle α is called the *angle of friction*. Up to the moment when the body slides, the resultant reaction of the plane on the body must evidently be along the vertical to balance W. The

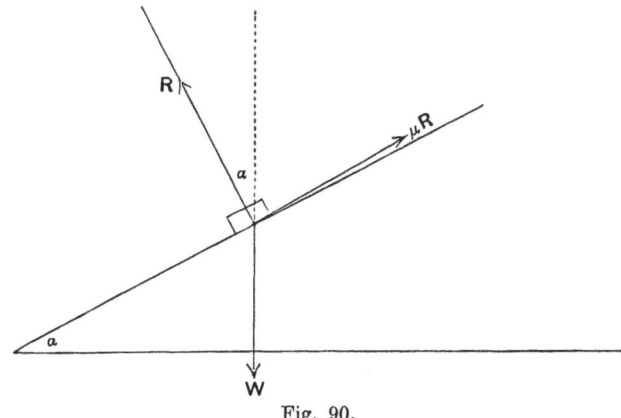

Fig. 90.

greatest angle the resultant reaction can make with the normal to the plane (when all the friction possible is called into play) is

$$\alpha = \tan^{-1} \mu.$$

Here, as well as in the method of § 186, it is better to start the body sliding and find the right slope of the plane to keep the velocity constant. For the friction between bodies that have been resting in contact for some time varies irregularly according to many other circumstances.

188. The treatment of problems when the friction is taken into account involves no new principles. But an extra force must be indicated in the figure, opposed to the direction in which motion is about to commence.

A very convenient way of including these forces is to draw *the cone of friction*, that is, the cone of angle

$$2\alpha = 2 \tan^{-1} \mu,$$

about the normal at the point of contact as axis. It is clear that the resultant reaction may lie anywhere inside this cone, but not out-

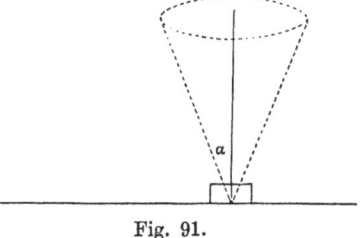

Fig. 91.

13—2

side of it. This consideration often serves to solve a problem by inspection.

For example, a beam is at rest but on the point of motion, inside a rough sphere, in a vertical plane through the centre. To find its inclination θ to the vertical.

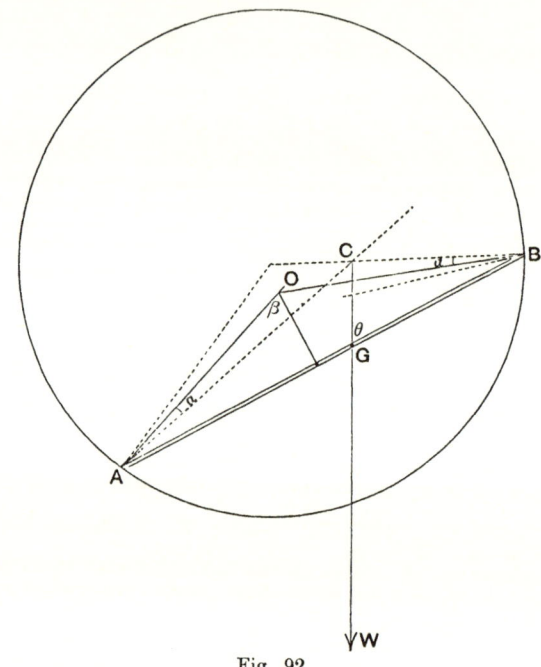

Fig. 92.

Let $\mu = \tan \alpha$ be the coefficient of friction for the beam and the sphere. Draw the cones of friction at A and B. Then since all the friction is called into play to prevent motion, the resultant reactions at A and B must lie along the edges AC, BC of the cones. The only other force acting on the beam is its weight. The three forces must meet in a point (§ 182), so that the vertical through C passes through the centre of gravity. Let the beam subtend an angle 2β at the centre of the circle.

Then

$$\angle ACG = \theta - \angle CAG = \theta - (90° - \beta - \alpha) = \alpha + \beta + \theta - 90°.$$
$$\angle BCG = 180° - \theta - \angle CBG = 180° - \theta - (90° - \beta + \alpha)$$
$$= 90° - (\alpha - \beta + \theta).$$

If the C.G. divide the beam in the ratio $m:n$, we see, by drawing horizontals through A and B to the vertical CGW, that

$$\frac{m}{n} = \frac{AC \sin ACG}{BC \sin BCG} = \frac{AC \cdot \cos(\alpha + \beta + \theta)}{BC \cdot \cos(\alpha - \beta + \theta)}$$

$$= \frac{\cos(\alpha - \beta)\cos(\alpha + \beta + \theta)}{\cos(\alpha + \beta)\cos(\alpha - \beta + \theta)},$$

whence θ may be found.

189. In the Simple Machines friction plays a very important part and cannot be neglected. In the case of the Screw and the Wedge its effects render the ordinary formulae practically useless. Recourse must then be had to direct experiment.

The student is strongly recommended to consult Sir Robert Ball's *Experimental Mechanics* for an admirable treatment of the principal machines with friction, with apparatus and weights on a practical scale. He will learn much besides from this book, as well the scientific method of dealing with sets of observations on variable quantities like friction, as the habit of keeping concrete facts in view when studying principles.

190. One general consideration may be mentioned. A part of the work done by the force applied to a machine is absorbed in overcoming the friction. It is converted into other forms of energy, *not destroyed*. But it is *lost for mechanical purposes*. The remainder is the useful work done by the machine.

Let E_l = the "lost" work.

E_u = the useful work.

P_1 = the force required to balance the weight when the machine works forwards.

P_2 = the force when the machine works backwards.

a = the distance moved by the "Power" handle.

Then
$$P_1 a = E_u + E_l,$$
$$P_2 a = E_u - E_l,$$

for when the machine works backwards, the friction aids the "Power."

Thus
$$(P_1 + P_2)\, a = 2E_u,$$

and
$$\frac{E_u}{P_1 a} = \frac{P_1 + P_2}{2P_1}.$$

This fraction which expresses the ratio of the useful work done to the total work done by the "Power" is called *the Efficiency* of the machine.

If the efficiency $=\frac{1}{2}$, P_2 is zero, so that no force is required to prevent the machine from running backwards, as friction is sufficient in itself to stop it; and this is *à fortiori* the case for smaller values of the efficiency.

For example, the Differential Pulley, as usually constructed, has an efficiency less than one half. The chain may thus be let go at any stage without risk of the machine running backwards. This useful property compensates for the waste of more than half the work done.

It is on this property that the usefulness of the *Wedge* depends; for only in virtue of friction can it be driven in by a series of blows, since otherwise it would slip back between each blow and the next.

This instrument is a double inclined plane. The "Power" is

The Wedge. applied parallel to the common base, and the resistance of the material to be divided acts as a pressure on the two faces.

Let 2α be the angle of the wedge; μ the coefficient of friction.

Then by resolving along the central line,

$$P = 2R \sin \alpha + 2\mu R \cos \alpha$$

and

$$\frac{P}{R} = 2(\sin \alpha + \mu \cos \alpha).$$

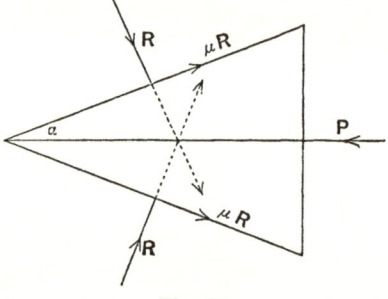

Fig. 93.

Apart from friction it is clear that by making α very small we may reduce this ratio as much as we please, so that a very small force will overcome a great resistance.

In cutting or piercing instruments the edges or points are

Theory of the Knife. ground down to an excessively small angle. The mechanical advantage may be still further increased by pushing or drawing the knife along, instead of pressing it in perpendicularly.

For if the pressure be applied in the direction $D'C$ (Fig. 94), inclined at an angle β to the edge of the knife, instead of along

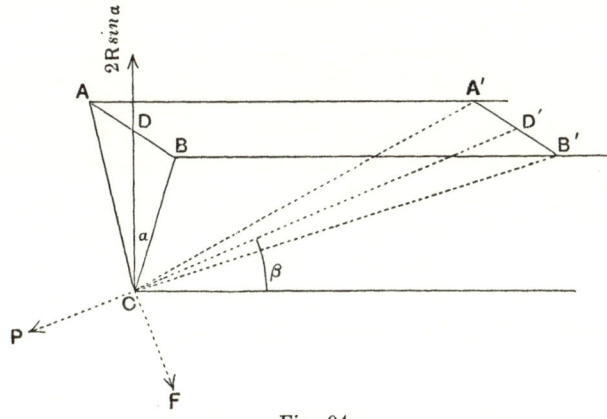

Fig. 94.

DC, the effective angle of the edge is $A'CB'$ instead of ACB, and this may be made as small as we choose by sufficiently inclining $D'C$.

If friction is neglected, the formula will in this case be

$$\frac{P}{R} = 2 \sin \alpha \cdot \sin \beta.$$

For the knife is supposed to be actuated by an oblique thrust P along $D'C$, and a pressure F at right angles to $D'C$, applied by smooth guides which compel it to travel in this direction, or by the hand itself. The forces F and $R \cos \alpha$ are at right angles to CD', hence resolving along CD' we obtain the above equation.

(Or by Lami's Theorem

$$\frac{P}{\sin \beta} = \frac{F}{\cos \beta} = \frac{2R \sin \alpha}{\sin 90°},$$

and therefore $\dfrac{P}{R} = 2 \sin \alpha \cdot \sin \beta$.

The same result follows from the principle of Work. For if the knife be moved obliquely along $D'C$ till it has sunk in to a vertical depth DC, no work is done by the pressure F, or by the forces $R \sin \alpha$, since their points of application remain at the same vertical height throughout. Hence the work done by P is equal to that done against the two thrusts $R \cos \alpha$. Now these forces have their points of application separated by a distance AB, so that the work done against them is $R \cos \alpha \cdot AB$.

Therefore $P \cdot CD' = R \cos a \cdot AB$,

an equation which at once leads to the formula

$$P = 2R \sin a \cdot \sin \beta.)$$

Another case of great practical importance is that of a rope wound round a rough post.

191.   Let a rope be passed round a rough circular post and
pulled at one end by a force T, while being held
Friction on a rope
wound round a   back by a force T_0 at the other.   We will in-
rough circular   vestigate the relation between T and T_0 in the
post.
case when T_0 is barely sufficient to prevent the
rope from slipping.   Let the part of the rope in contact with the
post subtend an angle α at the centre.

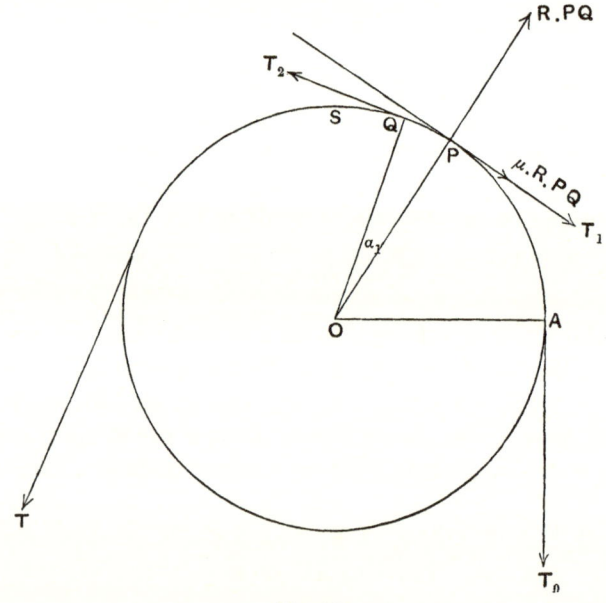

Fig. 95.

Consider a short length of the rope PQ subtending a small
angle $\alpha_1 = \dfrac{\alpha}{n}$ at the centre.

Let R be the normal pressure per unit length of the rope.
Then the pressure on PQ is $R \times PQ$, and the friction is $\mu R \cdot PQ$.
Let the tensions at P and Q be T_1 and T_2.

Resolving along the tangent and normal at P we have

$$T_2 . \cos \alpha_1 = T_1 + \mu R . PQ,$$
$$T_2 . \sin \alpha_1 = R . PQ.$$

Now ultimately, when α_1 is very small, $\cos \alpha_1 = 1$, and $\sin \alpha_1 = \alpha_1$. Also, if a is the radius of the post, $PQ = a\alpha_1$.

Thus $$T_2 = T_1 + \mu . T_2 \alpha_1.$$

In the term $\mu T_2 \alpha_1$ we may put T_1 for T_2, since the small difference between them, when multiplied by the small quantity α_1, will be negligible compared with the value of $T_1 \alpha_1$ itself. Hence

$$T_2 = T_1 (1 + \mu \alpha_1) = T_1 \left(1 + \frac{\mu a}{n} \right).$$

Let the tension at the end of the next short length QS, subtending α_1 at the centre, be T_3. Then, as above,

$$T_3 = T_2 \left(1 + \frac{\mu a}{n} \right) = T_1 \left(1 + \frac{\mu a}{n} \right)^2.$$

Proceeding in this way we see that if we start from a place where the tension is T_0 (say, where the rope first touches the post), the tension T at any other point at an angular distance α from the start is given by

$$T = T_0 \left(1 + \frac{\mu a}{n} \right)^n.$$

Writing $\dfrac{1}{x}$ for the small fraction $\dfrac{\mu a}{n}$, we have

$$T = T_0 \left(1 + \frac{1}{x} \right)^{x\mu a} = T_0 \left[\left(1 + \frac{1}{x} \right)^x \right]^{\mu a}.$$

If α_1 is taken very small, n and x become very large, and in the limit, for which alone the above reasoning is accurate, n and x become infinitely great.

Now by the Binomial Theorem

$$\left(1 + \frac{1}{x} \right)^x = 1 + x . \frac{1}{x} + \frac{x . \overline{x-1}}{1 . 2} . \frac{1}{x^2} + \frac{x . \overline{x-1} . \overline{x-2}}{1 . 2 . 3} . \frac{1}{x^3} + \dots$$

$$= 1 + 1 + \frac{1 - \dfrac{1}{x}}{1 . 2} + \frac{\left(1 - \dfrac{1}{x} \right) \left(1 - \dfrac{2}{x} \right)}{1 . 2 . 3} + \dots$$

$$= 1 + 1 + \frac{1}{1 . 2} + \frac{1}{1 . 2 . 3} + \dots$$

when x is infinitely large, since the part of each term neglected is infinitely small compared with what is retained.

It is proved in books on Algebra that this series is convergent, and, as more and more terms are taken, constantly approaches a finite limit between 2 and 3 in value. In fact it is the base of the natural system of logarithms, commonly denoted by e, and has the value

$$e = 2\cdot71828 \ldots.$$

Thus $$T = T_0 e^{\mu a}.$$

Hence the tension rapidly increases with α, and if the rope is taken two or three times round the post, an enormous force will be required to make it slip, although T_0, the force applied to prevent motion, may be very small. This is the principle employed in bringing boats to their moorings. A man with a few turns of rope round a post can hold up a great steamer, if the rope be strong enough.

192. The relation, discussed above, between the tension and the angle is worth careful consideration, for the student of Physics will meet with many similar cases. The essential feature is that while the angle increases in arithmetical progression the tension increases in geometrical progression. This is the relation between a power of a number and the corresponding index; or between a number and the corresponding logarithm; so that it may be described as the law of logarithmic growth.

The formula $$T = T_0 \left(1 + \frac{\mu\alpha}{n}\right)^n$$

is exactly analogous to that for the *amount* of a sum of money P_0 invested for n years at r per cent. compound interest,

$$P = P_0 \left(1 + \frac{r}{100}\right)^n.$$

So far, the increase takes place in a series of finite steps, corresponding to the addition of definite small angles in the one case, and to whole years in the other. In commerce the interest is sometimes added to the principal more frequently than by years—say quarterly. But in physics we have to go a step further, and suppose that the increment is instantly added to the principal as it accrues. The calculation of the last article shews that the formula then becomes

$$T = T_0 e^{\mu a}.$$

This law is met with whenever the rate of change of a quantity is *proportional to the amount of it already existing.*

In physics the law is more frequently found governing the rate of *decrease* at which some phenomenon dies away; as in the case of:

(1) The gradually decreasing swings of pendulums and galvanometer needles. The loss due to resistances is proportional to the length of swing over which they act, so that each swing is diminished by a definite fraction from the last. Hence in accurate work with the Ballistic Pendulum of § 269, it would be necessary to find what is called the *logarithmic decrement* by observing the proportional decrease in a given number of swings, and thus to allow for the effect of resistances in diminishing the first swing from its true value to what is observed.

(2) The discharge of a leyden jar, for the current is proportional to the remaining charge which drives it.

(3) The equalization of temperature by conduction, for the flow of heat is proportional at each instant to the outstanding difference of temperature.

(4) The progress of a chemical reaction between two substances, for the rate of change at each stage depends on the amount of uncombined reagents left.

EXAMPLES.

1. A block of iron weighing 56 lbs. rests on a stone floor, the coefficient of friction being ·3. What is the least force that will move it if applied (1) horizontally, (2) at 45° to the horizontal?

2. The base of an inclined plane is 4 feet and the height 3 feet. A force of 8 lbs. weight parallel to the plane will just prevent a mass of 20 lbs. from sliding down. Find the coefficient of friction.

3.   The weight on the driving wheels of a locomotive is 30 tons.   What is the greatest pull the engine can exert if the coefficient for iron on iron is ·2 ?

4.   A weight of 60 lbs. is on the point of motion down a rough inclined plane when supported by a force of 24 lbs. weight acting up the plane ; and is on the point of motion up the plane when the force is increased to 36 lbs. weight.   Find the coefficient of friction and the inclination of the plane.

5.   A mass of 80 lbs. rests on a plane ($\mu = $·15) inclined 30° to the horizon. Compare the forces required to draw it up the plane by a rope parallel to the plane and by a horizontal push.

6.   A heavy beam rests with one end on a rough floor and the other against a rough vertical wall, and is on the point of slipping when inclined 30° to the horizon.   Find μ, supposed to be the same for both surfaces.

7.   A ladder 20 feet long and weighing 70 lbs. rests at 45° against a rough vertical wall.   A man weighing 10 stone climbs up it.   If the coefficient of friction at each end is ·5, shew that it will slip when he has gone up 13 feet.

8.   Three complete turns of a rope are taken round a rough post, and one end of the rope is held with a force of 100 lbs. weight.   What force would be required at the other end in order to make it slip, if the coefficient of friction is ·3 ?

BOOK III.

APPLICATION TO VARIOUS PROBLEMS.

CHAPTER XX.

193. To find the motion of a body on a smooth inclined plane.

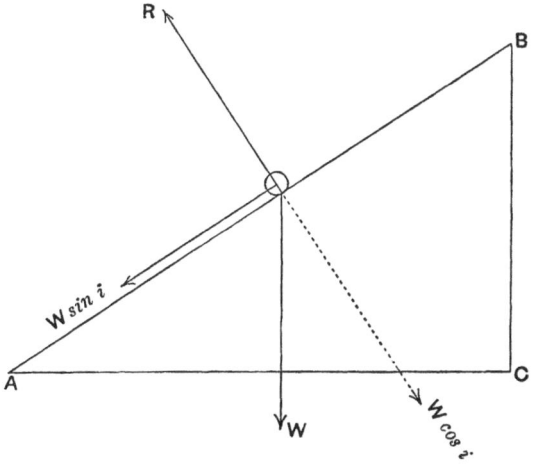

Fig. 96.

The weight W may be resolved into components $W \sin i$, down the plane, and $W \cos i$ perpendicular to it. Since there is no motion, and therefore no acceleration, perpendicular to the plane, the latter component must be just balanced by the pressure of the plane.

The weight W would produce an acceleration g in the body. The component $W \sin i$ produces the acceleration $g \sin i$ downwards. This value must be substituted for g in the kinematic formulae of § 111. The velocity acquired and the distance travelled in any time t from rest are given by

$$v = g \sin i . t,$$

$$s = \frac{g \sin i}{2} . t^2.$$

Also, if the body starts from rest at B, and v is its velocity when it arrives at A, then

$$v^2 = 2g \sin i . AB$$
$$= 2gAB . \sin i$$
$$= 2g . BC.$$

Thus the velocity at A is that due to the vertical fall BC, whatever the slope of the plane. (§ 66.)

194. The time of descent down all chords of a vertical circle terminating in the highest or lowest points is the same.

Let t be the time of descent down AC. The acceleration is $g \cos \theta$.

Then $AC = \dfrac{g \cos \theta . t^2}{2}$,

and $t = \sqrt{\dfrac{2AC}{g \cos \theta}} = \sqrt{\dfrac{2AB \cos \theta}{g \cos \theta}}$,

$\therefore t = \sqrt{\dfrac{2AB}{g}}$.

The time is obviously the same down the corresponding chord $C'B$ ending in B.

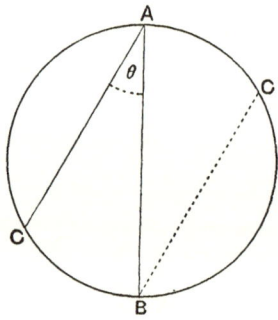

Fig. 97.

195. To find the shortest time of descent by an inclined plane from a given point to a given vertical circle.

Such paths of shortest time are called *brachistochrones*.

Could a circle be drawn to touch the given circle AB, and have its highest point at the given point P, the problem would be solved. For the time down PQ to Q, the point of contact, is obviously less than that down any other chord $PQ'R'$.

To describe the circle, join P to the lowest point of AB. This gives the path. For draw the vertical at P, and let the radius CQ produced cut it in O. Then

$$\angle OQP = \angle CQB = \angle CBQ = \angle QPO. \quad \therefore \quad OQ = OP.$$

Hence O is the centre of the circle required.

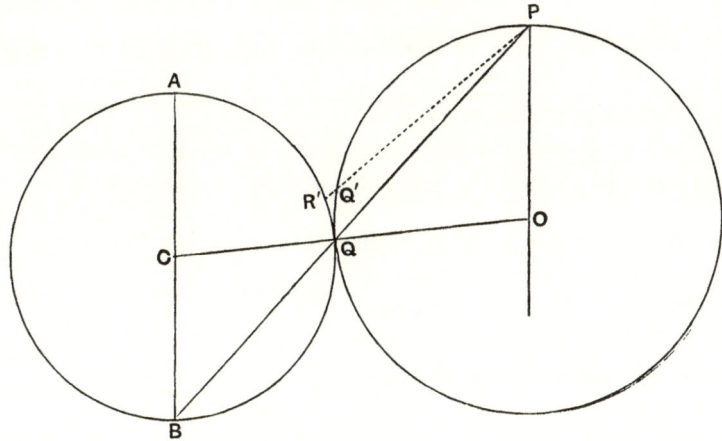

Fig. 98.

Similar problems about circles are generally solved by joining a point to the highest or lowest points of the circles.

196. The path of quickest descent between two curves bisects

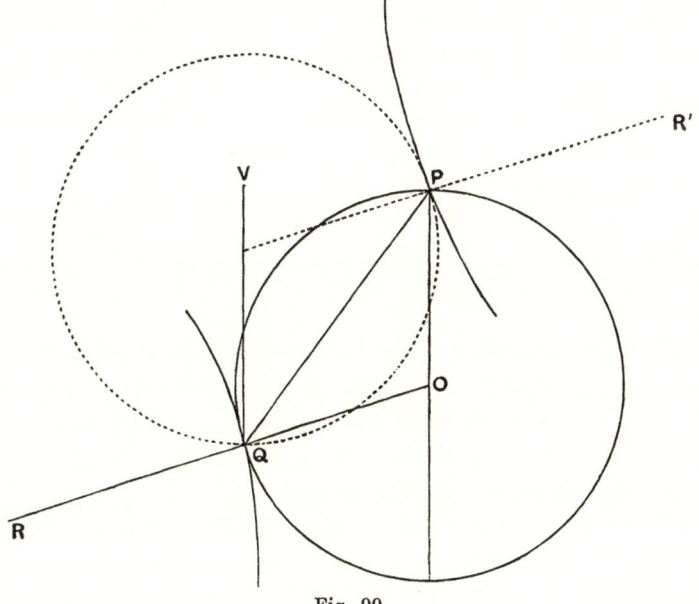

Fig. 99.

the angle between the vertical and the normal to the curve at each end.

If PQ is the path, P must be the highest point of a circle touching the lower curve at Q, OQ must be normal to the curve, and PQ obviously bisects OQV.

Similarly Q must be the lowest point of a circle touching the upper curve at P, and the property is true for that end also.

197. A curve may be regarded as the limiting form of a very

Motion on a curve under gravity. large number of inclined planes, joined end to end, when the number of planes is made infinitely great, and each plane infinitely short.

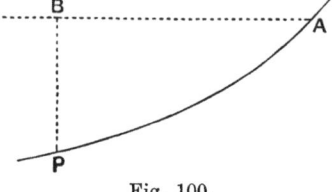

Fig. 100.

A body sliding down the curve acquires in each element of its path through the action of its weight the extra velocity due to the vertical height fallen through; and *in the limit* the pressure from the curve, which changes the *direction* of motion, is everywhere at right angles to the tangent, so that it can do no work which could change the magnitude of the velocity along the curve; so that the same law holds good for a finite length of the curve. The velocity acquired in sliding from A to P is the same as that acquired in falling from the vertical height BP. Galileo perceived this principle (§ 66), which is indeed the principle of Work. If m be the mass of the body, v the velocity at P, h the vertical height fallen through,

the energy $mv^2/2 = W \cdot h$ (the work done by the weight) $= mgh$,
and $v^2 = 2gh$.

198. Let a particle P be projected horizontally from A with

Motion on a vertical circle. velocity V, on a smooth circle APA'. To find the pressure on the curve at any point.

Let a be the radius of the circle; $AOP = \theta$.

(1) The velocity v at P is given by

$$\frac{v^2}{2} = \frac{V^2}{2} + g \cdot AN$$

$$= \frac{V^2}{2} + ga\,(1 - \cos\theta).$$

(2) At P the particle is describing a circle of radius a with velocity v. Its acceleration along the radius PO is $\dfrac{v^2}{a}$ (§ 77), depending only on the velocity at the moment.

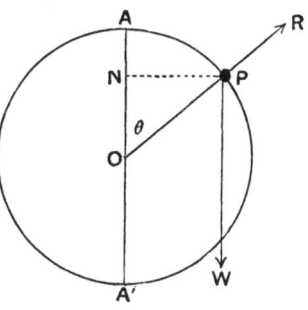

Fig. 101.

The resultant force acting on it along PO must be such as to give it this acceleration. The forces resolved along PO are $W \cos \theta - R$, where R is the outward pressure of the circle.

$$\therefore \frac{v^2}{a} = \frac{W \cos \theta - R}{m},$$

and

$$R = W \cos \theta - \frac{mv^2}{a}$$

$$= mg \cos \theta - \frac{m \left\{ V^2 + 2ga \left(1 - \cos \theta \right) \right\}}{a}$$

$$= m \left\{ (3g \cos \theta - 2g) - \frac{V^2}{a} \right\}.$$

If the particle slide from rest, $V = 0$ and

$$R = mg \left(3 \cos \theta - 2 \right).$$

This vanishes, *i.e.* the particle will leave the circle, when

$$3 \cos \theta - 2 = 0,$$

or

$$\cos \theta = \tfrac{2}{3}.$$

EXAMPLES.

1. A body is projected up a smooth plane inclined 30° to the horizon with a velocity of 80 feet per second. Find how far up the plane it will go before coming to rest, and the time occupied before it reaches the bottom again.

2. A toboggan on iron runners slides down an ice slope of 1 in 8 for a distance of 100 yards. Find the time occupied, and the speed at the bottom, neglecting friction.

3. Work question 2 allowing for a friction of $\mu = \cdot05$; and find how far the toboggan will then run on the level.

4. A heavy pendulum bob suspended by a wire 5 feet long is drawn aside 3 feet from the vertical and then let go. Find the velocity at the lowest point of the swing.

5. A boy on a swing 20 feet long works himself up till he can just touch with his feet a beam 4 feet below that from which the swing is hung. Shew that he must be going more than 20 miles an hour at the lowest point.

6. Give a geometrical construction for the line of quickest descent from a given point to a given inclined plane.

7. Find the shortest time in which a ring can be made to slide down a wire to a vertical wall from a point distant 20 feet from it.

8. Find the inclination of a plane down which a particle slides through a vertical height of 1 foot in half a second.

9. A one-ounce bullet, hung by a string 4 feet long from a fixed point, is projected so as just to describe a vertical circle without allowing the string to slacken when the bullet is at the top. Find the tension of the string when the bullet is at the bottom.

10. In a centrifugal railway the car runs down a long slope and then the rails take a turn round a vertical circle of 30 feet diameter, so that the car and passengers are upside down at the top. What is the least vertical height through which the car must descend before entering the circle in order that it may not leave the rails ?

11. A man is caught by the sail of a windmill, which is 29 feet long and revolves 10 times a minute. Shew that for a moment, just as he passes the top, he might let go his hold without falling.

CHAPTER XXI.

PROJECTILES.

199. GALILEO shewed (§ 74) that the path of a projectile in vacuo is a parabola.

One way of verifying this experimentally, when the particle is projected horizontally, is to allow a stream of water to issue in a horizontal jet from a supply at a steady pressure. The shadow of such a jet may be thrown by a distant candle on a sheet of ground glass, and traced with a pencil. The curve obtained will prove to be a parabola.

We shall describe a method devised by Morin to shew that if the motion of a body falling freely is compounded with a uniform horizontal velocity, the result is a parabola.

Experiment. Morin's Machine.

A sheet of paper is wrapped round a vertical cylinder which is turned uniformly on its axis by hand or by a motor. A small weight slides on a vertical rod at the side, and carries with it a pencil which presses against the paper. While the weight is held

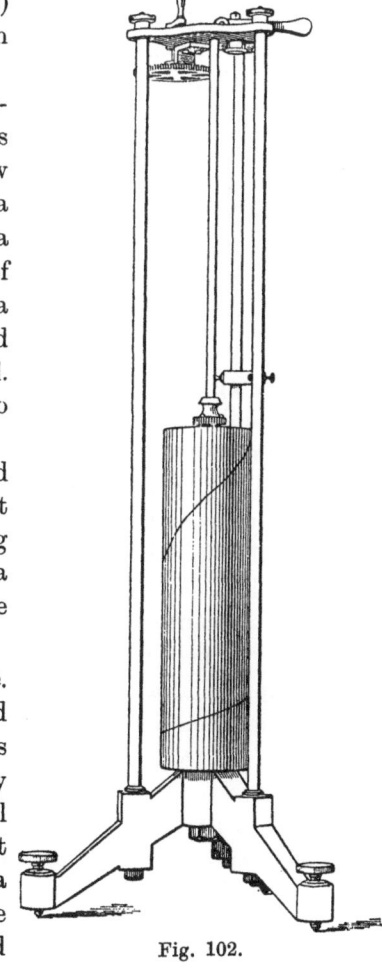

Fig. 102.

at the top of the rod, the pencil traces a horizontal circle on the paper with uniform speed. But when it is allowed to fall, it traces a curve resulting from the combination of the uniform horizontal motion of the paper with the motion of the freely falling weight. This curve proves to be a parabola.

The arrangement suffers from two defects. It is not easy to maintain absolute uniformity in turning the cylinder; and the friction of the pencil on the paper, and of the small weight on the rod, seriously affects the free movement of the weight. These defects are overcome in a form of machine designed by the Cambridge Scientific Instrument Company.

The pencil is in this case fixed on a vertical rod and can be adjusted so as to press against the paper at the bottom of the cylinder. The cylinder itself falls after being set in rapid rotation. It may be made very heavy, and yet brought to rest conveniently by a leaky piston arrangement at the end of the fall.

The rotation of the heavy cylinder is practically uniform throughout the short time of fall (cf. Galileo's Principle, § 70), since the very small frictions on the guiding rod and against the pencil are inconsiderable compared with the momentum of the heavy rotating cylinder; and if the rod is made truly vertical by means of the levelling screws, and is well oiled, the fall is practically "free" for the same reasons.

When the paper is unwrapped, a curve is obtained such as Fig. 104, inverted in the case of the second machine.

The only difficulty is to decide on the exact point A of the horizontal line, corresponding to the moment of release.

Choose a point A, and with a set square draw ANN' vertical, and mark off inches along it. At two divisions N, N', say 3 and 6 inches from A, draw horizontal lines NP, $N'P'$ to the curve, and measure their lengths carefully. Then we should find

$$\frac{PN^2}{AN} = \frac{P'N'^2}{AN'}.$$

If this is not very approximately correct, let us assume that the axis should have been drawn a short distance x to the left, as $A'nn'$. Then we should have

$$\frac{(PN + x)^2}{A'n} = \frac{(P'N' + x)^2}{A'n'}.$$

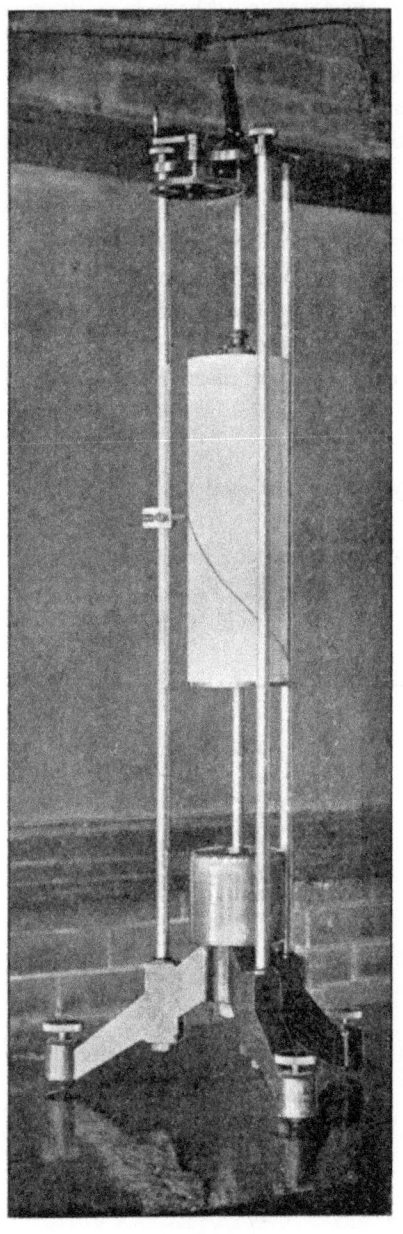

Fig. 103.

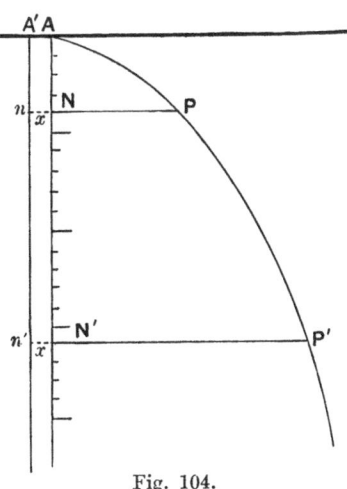

Fig. 104.

Solving, we find how far to the left the axis should have been drawn. If x turns out negative, the axis should be drawn to the right of AN. Having drawn the new axis, verify for several other distances AN'' that PN^2/AN is constant.

Let v be the horizontal velocity of the paper, t the time of fall.

Then $PN = vt$; $AN = gt^2/2$. So that PN^2/AN has the value $2v^2/g$. We may employ this result either to find the speed at which the paper was moving past the pencil horizontally, if we assume $g = 32 \cdot 2$; or to determine g if we have other means of finding the speed of the paper, as, for instance, by substituting for the pencil a tuning fork carrying a pointer after the manner of § 263 and using smoked paper.

200.   The proof of § 74 may be generalized for oblique pro-
Oblique Pro-   jection.
jection.   Let a body be projected in the direction PT with velocity u.

If gravity ceased to act, it would travel along PT with uniform velocity, and at the end of t seconds would arrive at T where

$$PT = ut.$$

If, on the other hand, it were merely dropped at P, it would in the same time t reach V where

$$PV = \frac{gt^2}{2}.$$

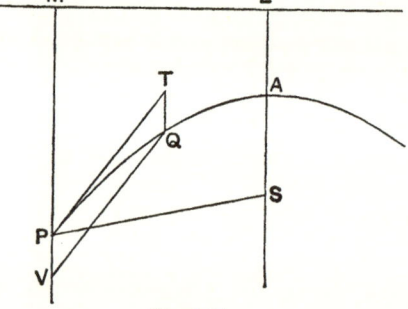

Fig. 105.

The initial velocity and gravity together bring it to Q, the other corner of the parallelogram on PT, PV.

By taking different times, t, we may construct any number of points Q on its path.

Draw the vertical PM, and make $PM = \dfrac{u^2}{2g}$.

On the other side of PT make $\angle TPS = \angle TPM$, and take

$PS = PM$. Draw ML horizontal, and with focus S and directrix SL describe a parabola. This is the path.

It passes through P, for $SP = PM$; and PT is a tangent, for it bisects SPM; PV is a diameter.

Also $QV^2 = PT^2 = u^2t^2 = \dfrac{2u^2}{g} \cdot \dfrac{gt^2}{2} = 4SP \cdot PV.$

Hence Q is on the parabola, and so is every other point similarly constructed.

201. The velocity at any point in the parabola is that due to a free fall from the directrix to that point.

For any point may be regarded as the point of projection. But the velocity due to a fall through MP is given by

$$\frac{v^2}{2} = g \cdot MP \quad (\S\ 111),$$

$$= g \cdot \frac{u^2}{2g},$$

so that $v = u$, the velocity at P.

202. To find the "elevation," or angle of projection, for a given velocity in order to strike a given point.

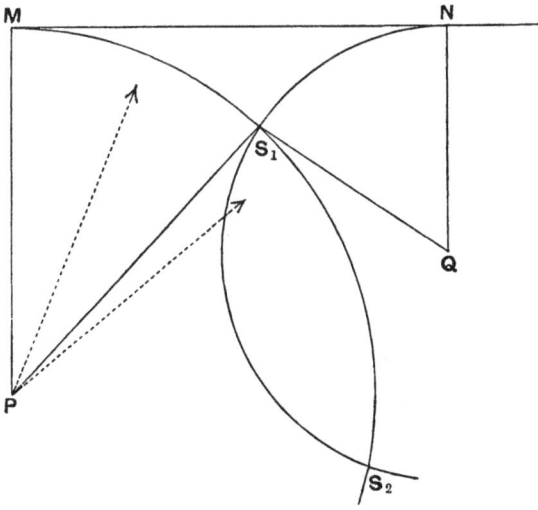

Fig. 106.

Let the body be projected from P with velocity V; let Q be the point to be struck.

Draw PM vertical and equal to $V^2/2g$. The horizontal MN will be the directrix, and P, Q are both on the parabola. Draw QN vertical, and describe the circles MS_1S_2, NS_1S_2 having P and Q as centres. Either point of intersection will do for a focus. Bisect the angles MPS_1, MPS_2. These are the directions of projection.

There are thus in general two possible angles of elevation for hitting a given object with a given velocity of projection. These reduce to one if the circles touch; if they do not intersect, the velocity is not great enough to reach Q.

203. The parabola helps us to realize the path as a whole, Analytical and problems may often be neatly solved by Method. means of its properties. But it is usually better to resolve the motion, and consider the horizontal and vertical components separately.

Let the body be projected with velocity u in a direction making an angle α with the horizontal.

Resolve u into a horizontal component $u \cos \alpha$, and a vertical component $u \sin \alpha$.

The only force acting on the body, once it is projected, is the weight, which causes a vertical acceleration g downwards. Thus we have

	Initial velocity.	*Acceleration.*
for the horizontal motion.........	$u \cos \alpha$,	0
„ vertical „	$u \sin \alpha$,	$-g$

(1) To find the velocity after t seconds.

The horizontal component $= u \cos \alpha$ ....(1).

(There is no acceleration to change it. Hence the horizontal velocity is constant; in order to keep exactly underneath a ball a cricketer must run with uniform speed.)

The vertical component $= u \sin \alpha - gt$ ....(2).

If v is the resultant velocity, inclined θ to the horizon,

$$v = \sqrt{u^2 \cos^2 \alpha + (u \sin \alpha - gt)^2}$$

$$\tan \theta = \frac{u \sin \alpha - gt}{u \cos \alpha}.$$

(2) To find the position at time t.

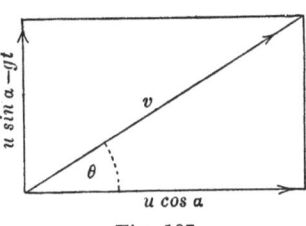

Fig. 107.

Horizontal distance $= u \cos \alpha . t$.

Vertical height $= u \sin \alpha . t - \dfrac{gt^2}{2}$.

(3) To find T the time of flight.

Gravity destroys a vertical velocity of g feet per second in every second. The body will be at the top of its path when gravity has just destroyed its upward velocity $u \sin \alpha$; *i.e.* in $u \sin \alpha / g$ seconds. It takes as long to come down. So the time of flight

$$= \frac{2u \sin \alpha}{g}.$$

(Or, by formula (2), at the top $u \sin \alpha - gt = 0$; hence

$$t = \frac{u \sin \alpha}{g}; \quad T = \frac{2u \sin \alpha}{g}.$$

Again, by formula (4), at the ground

$$u \sin \alpha . t - \frac{gt^2}{2} = 0,$$

$$T = 0 \text{ or } \frac{2u \sin \alpha}{g}.$$

The first value corresponds to the start, the second to the finish.)

(4) To find the range.

During the whole time of flight the horizontal velocity is constant. Hence the distance of the point where the projectile strikes

$$= \text{time of flight} \times \text{horizontal velocity}$$

$$= \frac{2u \sin \alpha}{g} \times u \cos \alpha$$

$$= \frac{u^2 \sin 2\alpha}{g}.$$

For a given velocity of projection this is greatest when $\sin 2\alpha = 1$; *i.e.* when $2\alpha = 90°$, and $\alpha = 45°$.

(5) To find the greatest height reached.

This is the same as for a body shot vertically upwards with velocity $u \sin \alpha$. Let h be the height. Then (§ 111),

$$\frac{u^2 \sin^2 \alpha}{2} = gh,$$

and
$$h = \frac{u^2 \sin^2 \alpha}{2g}.$$

(Or, by formula (4), height at time $\dfrac{u \sin \alpha}{g}$

$$= u \sin \alpha . \frac{u \sin \alpha}{g} - \frac{g}{2} . \frac{u^2 \sin^2 \alpha}{g^2}$$

$$= \frac{u^2 \sin^2 \alpha}{2g} .)$$

(6) To find the Latus Rectum of the parabola described.

At the top the vertical velocity has vanished, and only the horizontal component $u \cos \alpha$ is left. But this is the velocity due to a fall from the directrix to the vertex, *i.e.* through one quarter of the latus rectum. Hence the latus rectum

$$= 4 \times \frac{u^2 \cos^2 \alpha}{2g}$$

$$= \frac{2u^2 \cos^2 \alpha}{g}.$$

204. Problems relating to projection above an inclined plane

Projection over an Inclined Plane. are best solved by resolving along and perpendicular to the plane. Note the method only, which will be seen from a couple of examples.

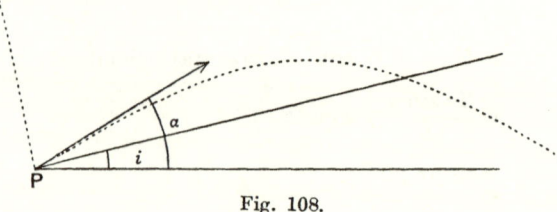

Fig. 108.

Ex. 1. A body is projected at elevation α with velocity u, from the foot of a hill of slope i; find the range on the hill.

Here the initial velocity has components

$$u \cos(\alpha - i) \text{ along the hill,}$$

and $\qquad u \sin(\alpha - i)$ perpendicular to it.

The acceleration of gravity has components

$$-g \sin i \text{ up the hill,}$$

and $\qquad -g \cos i$ perpendicular to it.

Velocity perpendicular to the hill at time t is

$$u \sin(\alpha - i) - g \cos i . t.$$

This vanishes when

$$t = \frac{u \sin(\alpha - i)}{g \cos i},$$

and the time of flight

$$= \frac{2u \sin(\alpha - i)}{g \cos i}.$$

Distance up the hill at time t

$$= u \cos(\alpha - i) . t - \frac{g \sin i}{2} . t^2.$$

This will be the range when

$$t = \frac{2u \sin(\alpha - i)}{g \cos i}.$$

$$\therefore \text{ Range} = \frac{2u^2 \sin(\alpha - i) \cos(\alpha - i)}{g \cos i} - \frac{g \sin i}{2} . \frac{4u^2 \sin^2(\alpha - i)}{g^2 \cos^2 i}$$

$$= \frac{2u^2}{g} . \frac{\sin(\alpha - i)}{\cos i} \left\{ \cos(\alpha - i) - \frac{\sin i . \sin(\alpha - i)}{\cos i} \right\}$$

$$= \frac{2u^2}{g} . \frac{\sin(\alpha - i) \cos \alpha}{\cos^2 i}.$$

Of course when $i = 0$ this reduces to the range on a horizontal plane

$$\frac{u^2 \sin 2\alpha}{g}.$$

Ex. 2. To find the condition that the projectile should strike the hill at right angles.

The velocity parallel to the hill at the moment of striking must be zero; *i.e.*, at $t = \dfrac{2u \sin (\alpha - i)}{g \cos i}$, we must have

$$u \cos (\alpha - i) - g \sin i \cdot t = 0,$$

$$\therefore \ u \cos (\alpha - i) - g \sin i \cdot \frac{2u \sin (\alpha - i)}{g \cos i} = 0.$$

$$\therefore \ \tan i = \tfrac{1}{2} \cot (\alpha - i),$$

which is the condition required.

EXAMPLES.

1. A ball is projected with a velocity of 80 feet per second at an elevation of 30° ; find the time of flight, range, and greatest height attained.

2. The muzzle velocity of a gun is 2240 feet per second. Find its greatest range in miles.

3. A shot is fired horizontally from a height of 4 feet above the ground, and strikes the ground at a distance of 1000 feet. What was its initial velocity ?

4. A man can throw a stone with a velocity of 80 feet per second. In what direction must he throw it so as to strike a bird 20 feet above his shoulder on a tree 30 feet away from him ?

5. A shot is fired from a fort 144 feet above the sea with a velocity of 2000 feet per second at an elevation of 30°. At what horizontal distance from the fort will it strike the water ?

6. The record throw for the cricket ball is 127 yards 1 foot 3 inches. Find the least initial velocity the ball could have received.

7. The record for the 16-lb. hammer is 171 feet. If the handle is 4 feet long, find how many times per minute it was being whirled round, assuming that it was let go in the most favourable position.

8. The back lines of a tennis court are 78 feet apart, and the service lines 42 feet. The net is 3 ft. 3 in. high.

Find the horizontal velocity of the ball

(1) when it is returned from near the ground at one back line so as to graze the net and just strike the other back line ;

(2) when it is served from a height of 8 feet, grazes the net, and strikes the service line.

9. A bullet is fired from a rifle 4 ft. above the ground with a velocity of 2000 feet per second, so as to strike a target at 500 yards at the same height of 4 feet. Shew that the rifle must be sighted for an elevation of about 20′ ; and that a man 6 ft. 2 in. in height could stand halfway between the rifle and target without being hit.

10. The record for the high jump is 6 ft. 5⅜ in. What is the greatest upward velocity a man can give his body in leaping, assuming that his centre of gravity is 3 ft 6 in. above the ground and that it rose 15 inches above the bar ?

What height could the same man jump at the surface of the moon, where gravity is 150 c.g.s. units ?

11. A shot is projected with a velocity of 2000 feet per second at an angle of 30° with a hill which is itself inclined 30° to the horizon. Find the time of flight and range up the hill.

12. The greatest range of a gun on a horizontal plane is 3000 feet. Shew that its greatest ranges up and down a plane inclined 30° to the horizon are 2000 feet and 2000 yards respectively.

13. The record for putting the weight (16 lbs.) is 48 ft. 2 in. Assuming that it is let go at a height of 6 feet above the ground, and elevation 45°, calculate the speed with which it is projected. If in the act of projection it is heaved up through a vertical height of 2 feet, find the foot-pounds of work put into it.

14. The Norwegian Ski-jumping contests in February, 1904, took place on a snow slope at Holmenkollen, 186 yards long. The competitors slid down two-thirds of the slope (which was in this part inclined 15° to the horizon) to a ledge, from which they took off for the jump. Below the ledge the steepness of the slope increased to 24°. Supposing that the coefficient of friction between the ski and the ice was ·05, and that the lip of the ledge was so curved as to give the jumper an elevation of 6° above the horizon at the take-off, shew that the speed at the ledge would be about 48 miles an hour, and the leap about 104 feet.

CHAPTER XXII.

SIMPLE HARMONIC MOTION.

205. *Definition.* Let a point P describe a circle with constant velocity. Draw PN perpendicular to the diameter AOA'. The motion of N is called a Simple Harmonic Motion (s.h.m.).

Let the diameter AOA' be horizontal.

As P revolves constantly round the circle, N vibrates or oscillates about the centre O, between the extreme positions A and
A'. At A the motion of P is
entirely vertical, and N is for an
instant at rest. It then gains
speed towards O, until at O it has
momentarily the full speed of P,
moving parallel to AOA' at B
overhead. Next N slackens pace
till it comes to rest for an instant
at A'; reverses its motion; and at
O has the full speed of P to the
right; and finally slackens again
till it comes to rest at A as P

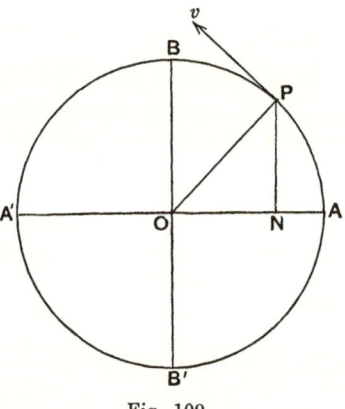

Fig. 109.

passes through. The motion is repeated with every revolution
of P.

This kind of motion is extremely important. It may almost
be said to share the whole range of Physics with the law of
gravitation; for whenever we are not dealing with the mutual
action of distant bodies by means of the one, we shall almost

certainly find ourselves working out the movements of connected bodies or parts of bodies by the other.

The simplest way to study it is to observe that the point N is subject to the horizontal motion of P without any of its vertical motion. We know the velocity of P; and its acceleration has been found (§ 77). The velocity and acceleration of N at any instant will be the horizontal components of the velocity and acceleration of P at that instant.

But first we will give another investigation of the acceleration of P partly for the sake of the importance of the result; partly, too, for the interest of the graphical method it illustrates, that of the *Hodograph.*

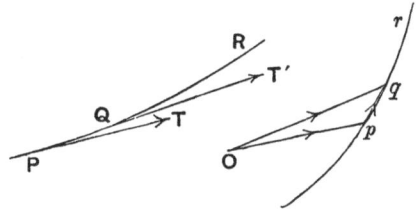

Fig. 110.

206. *Definition.* Let a point describe any curve PQR in any manner. Make a construction diagram of its velocity. From any point O draw Op to represent in magnitude and direction its velocity at P; Oq to represent its velocity at Q, and so on. For successive points on PQR there will be corresponding points pqr through which a curve may be drawn. This curve is called the Hodograph of the orbit PQR.

The Hodograph.

Property of the Hodograph.

The velocity in the hodograph represents the acceleration at the corresponding point of the orbit.

For, by the triangle of velocities, the straight line pq represents the velocity which must be compounded with Op to produce Oq; i.e., pq is the velocity gained by the point in passing from P to Q in the orbit, while, if q is taken very near to p, the line pq becomes the arc of the hodograph.

C.　　　　　　　　　　　　　　　　　　　15

The rate at which pq is described is thus the rate of change of velocity in the orbit, and also the rate of description of arc in the hodograph.

Hence the *velocity* in the hodograph is the *acceleration* in the orbit.

The advantage of the hodograph is that it reduces the study of accelerations to that of velocities, usually less complex.

207. Let P describe a circle, centre O, radius $OP = r$, with velocity v.

Acceleration of a point which describes a circle with uniform velocity.

Take O for origin of the hodograph. Draw Op representing v, the velocity at P, and therefore parallel to the tangent PT. As P moves round O, the line Op revolves with constant length. Thus the hodograph is a circle, and the velocity of p is at right angles to Op, *i.e.*, parallel to PO.

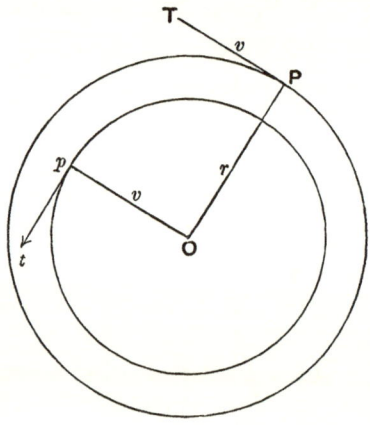

Fig. 111.

The acceleration of P is therefore along PO, *i.e.*, towards the centre. And since the circles are described in equal times,

$$\frac{\text{Velocity of } p}{\text{Velocity of } P} = \frac{\text{radius } Op}{\text{radius } OP}.$$

$$\therefore \quad \frac{\text{Acceleration of } P}{v} = \frac{v}{r},$$

and acceleration of $P = \dfrac{v^2}{r}$.

As in § 77, we may write

$$\text{acceleration of } P = \frac{4\pi^2}{T^2} \cdot r,$$

where T is the time of one complete revolution, or Periodic Time of P.

208. To find the Acceleration in a S.H.M.

Let T be the periodic time of P.

The acceleration of P is $\dfrac{4\pi^2}{T^2} OP$ along PO.

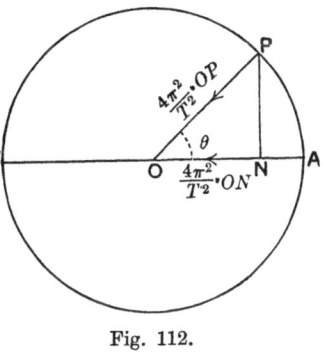

Fig. 112.

The acceleration of N = horizontal component of acceleration of P

$$= \frac{4\pi^2}{T^2} PO \times \cos\theta,$$

$$\frac{4\pi^2}{T^2} . NO$$

$$= -\frac{4\pi^2}{T^2} . ON.$$

Note that (1) the acceleration is always directed towards O, for if N is to the left of O, ON is negative, and the acceleration becomes positive; and (2) its magnitude is proportional to ON at every point, for the coefficient $4\pi^2/T^2$ is the same throughout the vibration.   This is no longer a constant acceleration, like that of gravity for falling bodies and projectiles, but an acceleration proportional to the distance from a fixed point, and always directed towards it.

The acceleration is greatest at A and A', where the velocity is zero (but is being reversed in direction); and vanishes at O, where the velocity is greatest (and for a moment remains equal to that of P; hence no change of velocity, *i.e.*, no acceleration).

209. The force required to produce such an acceleration in any given mass must, by the Second Law of Motion, be proportional to the distance of the mass from the fixed point O.   This is precisely the law that is found to hold good when elastic bodies are pulled or twisted out of shape; and in general for any disturbance of stable equilibrium among systems of material bodies, or the parts of the media that transmit sound, and the phenomena of light, electricity, and magnetism, at all events if the disturbances are small.   The forces that are at once called into play tending to restore the normal condition of equilibrium are, for small disturbances, proportional to the displacements. Hence the various systems or their parts begin a series of Simple Harmonic Vibrations about their positions of equilibrium. Even if the initial circumstances lead to more complicated forms of vibration, Fourier shewed (*Théorie de la Chaleur*) how to resolve any form of oscillation into a series of Simple Harmonic Vibrations.

210. 1. Suspend a small scale

Experimental
Verification.
pan (1) by an elastic thread, (2) by a spiral spring, in front of a vertical scale, and read its position. Add weights and take readings of the position of the pan. It will be found that the amount of descent is proportional to the weight added.

The principle is employed in Joly's Balance for small weights, and in the ordinary spring balances for large ones.

2. Support a square rod of iron or wood horizontally across two knife edges on a lathe bed. Attach a vertical millimeter scale to the centre of the rod, and hang various weights by a hook at the centre. The deflections can be read by a fixed microscope furnished with a cross-hair, and directed upon the millimeter scale. They will be found to be proportional to the weights employed.

3. Suspend a heavy cylinder, furnished with a pointer moving over a horizontal circular scale, by a brass wire hanging from a

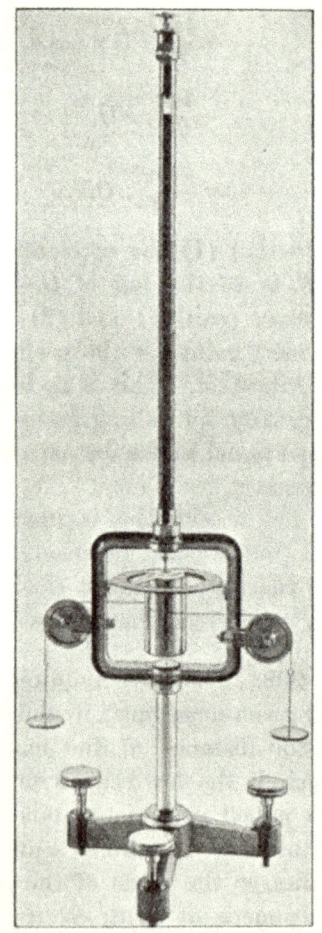

Fig. 113.

fixed clamp. Threads attached to the rim of the cylinder, and wound round it, leave it at opposite ends of a diameter, and passing over pulleys support equal small weights. The angle through which the cylinder turns may be read by graduations on the rim, and is the angle through which the wire is twisted. Verify that it is proportional to the weight used, and therefore to the moment or torque of the twisting couple.

Thus for extension, flexure and torsion the displacement is proportional to the force required to produce it; so that the force tending to restore the normal condition is proportional to the displacement.

211. *Def.* 1.   The greatest distance of displacement from the position of equilibrium is called the *Amplitude* of the vibration.

Amplitude—Periodic Time—Phase—Epoch.

Def. 2.   The time of one complete vibration, *i.e.*, between two passages through the same point in the same direction, is called the *Period*. This is the same as the time of one revolution of P in the circle. The period or periodic time, means the time of a double swing, say from A to A' and back again. There is one exception to this rule; in all work relating to the measurement of time it has become the custom to define the " seconds" pendulum as one which executes a half oscillation in one second, that is which swings from A to A' in one second.

Def. 3.   The position of the point N in its vibration is called the *Phase*. It is measured by the fraction of a vibration which has taken place since the beginning of the vibration.

Let t be the time elapsed since the beginning of the vibration at A; θ the angle described by OP. Then the phase is indicated by t/T or $\theta/2\pi$. Differences of phase are often expressed as differences of the angle θ, directly in degrees.

Def. 4.   If the time is

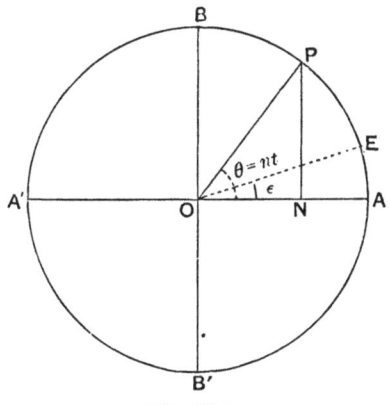

Fig. 114.

measured, not from the moment when P passes through A, but from some other position of P, as E, then the angle AOE is called the *Epoch*, and is generally denoted by ϵ.

212. Everything about a s.h.m. can now be easily expressed.

Let the amplitude (equal to the radius of the circle of reference) be r; $ON = x$; $n =$ the angular velocity of OP (more usually denoted by ω except in this connection).

Then the position, velocity, and acceleration of N are given by

$$x = r \cos \theta = r \cos \frac{2\pi}{T} . t = r \cos nt,$$

$v =$ component of P's velocity, which is $2\pi r/T$,

$$= - \frac{2\pi}{T} . r \sin \theta = - nr \sin \theta = - nr \sin nt.$$

$$a = - \frac{4\pi^2}{T^2} . r \cos \theta = - n^2 r \cos \theta = - n^2 r \cos nt$$

$$= - n^2 x.$$

If the time is measured from E, we have only to add the epoch to each of the angles.   Thus $x = r \cos (nt + \epsilon)$.

213. When a point receives displacements (and consequently velocities and accelerations) which are at every instant the resultants of those due to two s.h.m.'s, its motion is said to be compounded of the two

Composition of Simple Harmonic Motions.

s.h.m.'s.

In 1821 Fresnel ("Mémoire sur la Diffraction," *Comptes Rendus*, 1826) gave a rule for compounding s.h.m.'s of the same period and in the same straight line.

Let the parallelogram $OACB$ revolve uniformly round the vertex O, and drop perpendiculars AP, BQ, CR on a fixed line OX. Then P, Q, R all describe s.h.m.'s along OX. Also

$$OR = \text{projection of } OC \text{ on } OX$$

$$= \text{sum of projections of } OA, AC$$

$$= \text{sum of projections of } OA, OB$$

$$= OP + OQ.$$

Thus the displacement of R is always the sum of the displacements of P and Q. Now the amplitudes of the three vibrations are OA, OB, OC; and their phases at any moment are fixed by the angles XOA, XOB, XOC.

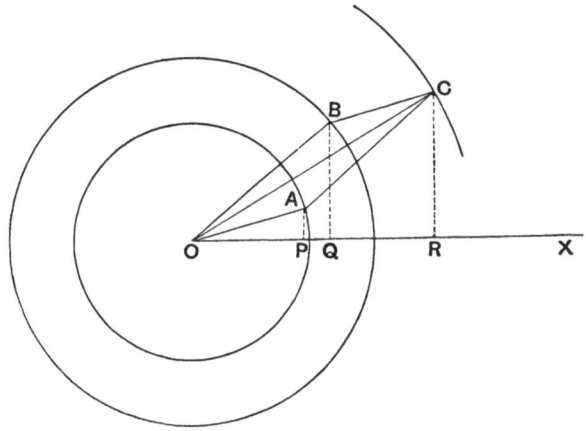

Fig. 115.

It appears that *the resultant of two* S.H.M.*'s of the same period and in the same straight line is another* S.H.M. *in that line with different amplitude and phase, which are to be found by a construction diagram precisely as the magnitude and direction of the resultant of two forces is found in Statics.*

This is known as Fresnel's Rule. Thus, from any point O draw OA to represent the amplitude of the first vibration at such an angle with a fixed line OX as will represent its phase. From A draw AC to represent in the same way the amplitude and phase of the second vibration. Join OC. Then OC represents the amplitude and phase of the resultant vibration.

This corresponds to the Triangle of Forces (§ 166), and it may obviously be extended to find the resultant of any number of S.H.M.'s, as in the Polygon of Forces (§ 168).

The result is extremely important in the theories of Sound, Light, and Electricity, where the vibration at any point is generally the resultant of many disturbances received simultaneously from different parts of an advancing wave.

Corollary 1. If the component vibrations are in the same phase, OAC is a straight line, and the amplitude of the resultant is the sum of the amplitudes of the components. If they differ by half a period, the resultant amplitude is the difference of the amplitudes of the components. In the latter case if the amplitudes of the components are equal, the resultant is zero; so that two equal S.H.M.'s differing in phase by half a period destroy each other.

Corollary 2. If the periods are very nearly, but not quite, equal, one vibration will gradually gain in phase on the other, and the effect will be the same as if the periods were exactly equal, but the difference of phase were gradually increased. The resultant will be a vibration with changing amplitude, which varies between a maximum, equal to the sum of the original amplitudes, and a minimum equal to their difference.

Illustrations of this effect are found in the theories of Interference of Sound and Light, and of alternating currents in Electricity.

214. Two S.H.M.'s in different directions generally compound into motion in a curve. By way of example, we will consider two important cases of the composition of equal S.H.M.'s of the same period in directions at right angles to each other.

(1) When the phase is the same.

Let the vibrations start from O along OA, OB at the same epoch. Then the distances along OA, OB are the same at all times for both. A point having both these vibrations would move along the diagonal OA'', and describe a S.H.M. of the same period, but of amplitude

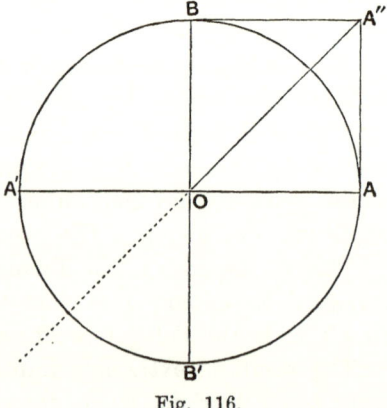

Fig. 116.

$$OA'' = \sqrt{\overline{OA^2 + OB^2}}.$$

(2)   When the phase-difference is 90°, or one quarter of a period.

In the figure of § 205 let PM be drawn perpendicular to the diameter BOB'. The motion of M in BOB' is precisely the same as that of N in AOA', but M starts from O when N is already at A, *i.e.*, the motion of M is a quarter of a period, or 90° in phase, behind that of N. The motion of P is evidently the resultant of the motions of N and M. Hence:

Two S.H.M.*'s of equal amplitude and period, but differing in phase by a quarter of a period, or 90°, compound into uniform motion in a circle.*

If M had been a quarter of a period in front of N, the motion in the circle would have been in the opposite direction.

(3)   It is sometimes useful to resolve a S.H.M. into two uniform circular motions in opposite directions.

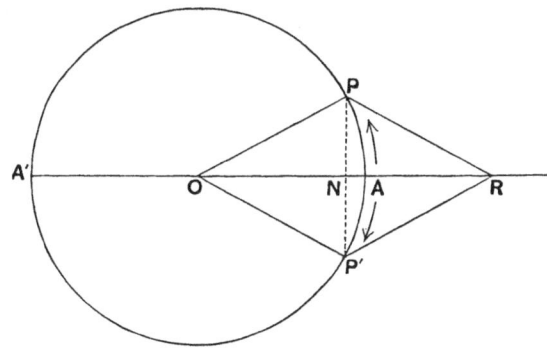

Fig. 117.

Let two points P, P' start from A and describe a circle uniformly in opposite directions. The two circular motions are together equivalent to two S.H.M.'s along AOA', and two others at right angles to AOA'. The latter, differing in phase by half a period, destroy each other. The resultant is therefore a S.H.M. along AOA' with amplitude *double* the radius of the circle.

Conversely a S.H.M. may be resolved into two uniform circular motions in opposite directions, the radius of the circle being *half* the amplitude of the vibration.

This may be seen directly from the figure, for the resultant of the displacements OP, OP' is OR, equal to twice ON. Such a motion is realized by the arrangement described in *Example* 5, below.

Applications of these results are found in the theory of polarized light.

215. Whenever the force acting on any mass is proportional to the distance from a fixed point, and directed towards it, so also will the acceleration be, and the motion will consequently be Simple Harmonic Motion. Suppose we know, from a consideration of the forces acting, that in some given case the ratio of the acceleration to the distance ON is μ; so that for this case

$$a = -\mu . ON.$$

We have proved (§ 208) that if T is the periodic time of the vibration,

$$a = -\frac{4\pi^2}{T^2} . ON.$$

Hence
$$\frac{4\pi^2}{T^2} = \mu,$$

and
$$T = \frac{2\pi}{\sqrt{\mu}}.$$

This does not depend on the amplitude or distance of displacement, but only on the ratio $\dfrac{acceleration}{distance}$.

Such vibrations, whose period, or time of swing, does not depend on the amplitude, are called *isochronous*, *i.e.*, executed in the same time.

Experiment. Verify this property with the apparatus of experiments of 1 and 3 of § 210.

EXAMPLES.

1. In a s.h.m., find the time of an oscillation

(1) when the acceleration at a distance of 3 inches from the centre is 4 ft.-sec. units;

(2) when the acceleration at a distance of 25 centimeters from the centre is 625 cm.-sec. units;

(3) when the force acting on the body at a distance of 2 feet is equal to its weight.

2. Find the velocity at the centre in each of the above cases when the amplitudes are respectively (1) 1 foot, (2) 10 cm., (3) 2 feet.

3. Shew that in a s.h.m. of amplitude a, and periodic time $\dfrac{2\pi}{\sqrt{\mu}}$, at a distance x from the centre the velocity is $\sqrt{\mu\,(a^2 - x^2)}$, and the time elapsed since it was at its greatest distance $= \dfrac{1}{\sqrt{\mu}} \cos^{-1} \dfrac{x}{a}$.

4. In a s.h.m. the velocities at distances 5 and 12 feet are 36 and 15 feet per second respectively; find its period and the acceleration at the greatest distance from the centre.

5. One end of a rod 2 feet long is pivoted on a pin projecting from the edge of the rim of a flywheel 4 feet in diameter. The other end, which carries a shelf, is attached to a block which runs up and down a vertical slot passing through the line of the wheel's axis. Shew that the motion of the shelf is a s.h.m. What is the greatest number of revolutions per minute the wheel can make without causing objects placed on the shelf to part company with it at the top?

6. Hooke discovered that the tension of an elastic thread is proportional to its extension per unit length; so that if l is the unstretched length, the tension T required to stretch it to a length l' is, according to Hooke's law,

$$T = T_0 \cdot \frac{l' - l}{l},$$

where T_0 is a constant which is evidently the tension required to stretch the thread to double its length.

A thread 30 cms. long, which when supporting 5 gms. extends to a length of 50 cms., is fastened 30 cms. below a small hole in a smooth table. The other end is drawn up through the hole and attached to a gramme weight. The weight is placed on the table at a distance of 10 cms. from the hole, and let go. Find the time of an oscillation, and the velocity of the weight as it passes over the hole.

7. A brass clamp weighing 10 gms. is fastened to one prong of a fixed vertical tuning fork, and when the fork is excited by a violin bow, the clamp vibrates 256 times per second through a distance of 1 mm. Calculate the force (in dynes) exerted on the clamp at the extremity of a vibration.

8. In a s.h.m. of amplitude 10 feet and period 15 seconds find the time occupied in travelling (1) 5 feet, (2) 7 feet from a point of rest.

CHAPTER XXIII.

THE SIMPLE PENDULUM.

216. THE most familiar case of a S.H.M. is the Simple Pendulum. This consists of a bob of mass m hanging from a fixed point by a string. Theoretically the mass of the bob should be collected at a point, and the string should have no weight or mass.

Let l be the length of the string; θ the angle it makes with the vertical at any instant.

The forces acting on the bob are the weight W vertically downwards, and the tension of the string T.

Resolving along the tangent at P we see that the only force affecting motion along the circle is the component of W,

$$W \cos (90° - \theta) = W \sin \theta,$$

directed towards O; since the tension T has no component along the tangent.

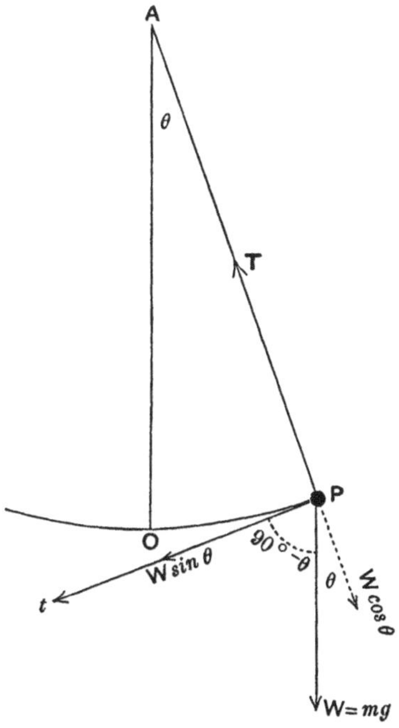

The acceleration is therefore

$$= - \frac{W \sin \theta}{m}$$

$$= - g \sin \theta.$$

Fig. 118.

If θ is very small, we may put the circular measure for the sine, so that

$$a = -g \cdot \frac{\text{arc } OP}{\text{radius } AO}$$

$$= -\frac{g}{l} \cdot OP.$$

The acceleration is thus proportional to the displacement, since neither g nor l depends on the displacement.

But this is the law for s.h.m.,

$$a = -\frac{4\pi^2}{T^2} \cdot ON.$$

Therefore the pendulum describes simple harmonic vibrations of a period T, such that

$$\frac{4\pi^2}{T^2} = \frac{g}{l},$$

and

$$T = 2\pi \sqrt{\frac{l}{g}}.$$

Observe, this is only true for small angular vibrations of a few degrees on each side of the vertical. *For small angles the swings of a pendulum are isochronous.*

217. *Experiment.* Suspend two leaden or iron balls by threads of equal length from a lofty ceiling or gallery. Draw one of them aside a few inches, and with the other hand draw the other as far behind you as you can reach. Release them exactly at the same instant, by opening the thumbs and fingers so as not to give either of them the slightest push. Watch the first pendulum, and at the moment when it returns to your hand close the other hand without looking round. *It will grasp the other pendulum.*

This principle is said to have been first detected by Galileo who timed the decreasing swings of a bronze lamp (a masterpiece of Benvenuto Cellini) during a service in the cathedral of Pisa by counting the beats of his pulse.

Huyghens, who rendered great services in the invention and perfection of the pendulum clock, discovered the cycloid and how to make a pendulum swing in it. In this curve the vibrations are strictly isochronous whatever their amplitude. But in practice it is better to use the simple circular pendulum and keep the angular vibrations small.

218. For a seconds pendulum the time of a single swing, *i.e.*, a half period, must be one second. Hence

Length of the
Seconds
Pendulum.

$$\pi\sqrt{\frac{l}{g}} = 1,$$

and

$$l = \frac{g}{\pi^2}.$$

At Greenwich $g = 32\cdot1912$ and consequently $l = 39\cdot13983$ inches.

219. Let n be the number of vibrations with length l; n' the number lost for a length $l + \lambda$.

Number of vibra-
tions lost in a day
for a small increase
in the length.

Then since the number of vibrations in a given time varies inversely as the periodic time,

$$\frac{n - n'}{n} = \sqrt{\frac{l}{l + \lambda}},$$

or,

$$1 - \frac{n'}{n} = \frac{1}{\left(1 + \dfrac{\lambda}{l}\right)^{\frac{1}{2}}} = \left(1 + \frac{\lambda}{l}\right)^{-\frac{1}{2}} = 1 - \frac{\lambda}{2l},$$

neglecting squares of $\dfrac{\lambda}{l}$.

$$\therefore\ n' = \frac{\lambda}{2l}\cdot n = \frac{\lambda}{2l}\cdot86400,$$

since a day of 24 hours contains 86400 seconds.

220. Let γ be the increase in g, n' the number gained.

Number of vibra-
tions gained for a
small increase in
the value of *g*.

As before ;

$$\frac{n + n'}{n} = \sqrt{\frac{g + \gamma}{g}} = \left(1 + \frac{\gamma}{g}\right)^{\frac{1}{2}}.$$

$$\therefore\ 1 + \frac{n'}{n} = 1 + \frac{\gamma}{2g} + \ldots\ldots$$

$$\therefore\ n' = \frac{\gamma}{2g}\cdot n.$$

221. For a point outside the earth gravity varies inversely as the square of the distance. (§ 238.)

At a height h, g is less by γ where

$$\frac{g - \gamma}{g} = \frac{R^2}{(R + h)^2} = \frac{1}{\left(1 + \dfrac{h}{R}\right)^2} = 1 - \frac{2h}{R} + \ldots\ldots,$$

R being the earth's radius.

$$\therefore \frac{\gamma}{g} = \frac{2h}{R}.$$

But

$$n' = \frac{\gamma}{2g} . n = \frac{h}{R} . n.$$

$$\therefore h = \frac{n'}{n} . R.$$

222. For a point within the earth gravity varies as the distance from the centre (§ 239). Hence at depth d,

$$\frac{g - \gamma}{g} = \frac{R - d}{R}.$$

$$\therefore \frac{\gamma}{g} = \frac{d}{R},$$

$$\frac{n'}{n} = \frac{\gamma}{2g} = \frac{d}{2R}.$$

$$\therefore d = \frac{n'}{2n} . R.$$

223. The formula $\quad T = 2\pi\sqrt{\dfrac{l}{g}}$

can be applied to determine the value of gravity.

For $\quad g = \dfrac{4\pi^2}{T^2} . l.$

It is far easier to obtain accurate values of the length of a pendulum, and its time of swing, than to observe the distance traversed in one second by a falling body; and if the velocity of the falling body is diminished by some device such as the Atwood's Machine, errors are introduced by the various frictions which more than compensate for the greater ease of observation. Newton employed this method, and using as bobs for his pendulums equal boxes filled with different materials, he was able to shew with great accuracy that the value of gravity was the same for all substances at the same place.

Experiment. A leaden sphere is hung by a long string, and set swinging through a small angle.

Take the time of 50 or 100 swings and find the time T of one complete double swing.

Measure the length of the string by a steel tape; and add half the diameter of the sphere as determined by the calipers. Call this l.

Then
$$g = \frac{4\pi^2}{T^2} \cdot l.$$

In this experiment the weight of the string has been neglected; the mass of the bob has been taken as collected at the C.G.; and it is not easy to observe with extreme accuracy either the distance from the point of suspension to the C.G., or the time of swing. These difficulties are overcome by the use of Kater's Pendulum (§ 276).

EXAMPLES.

1. What is the length of a pendulum beating half-seconds at Greenwich?

2. The bob of a seconds pendulum is screwed up 1/32 of an inch. How many seconds will it gain in a day?

3. A clock with a seconds pendulum is found to be gaining 7 seconds per day. How much must the bob be let down?

4. Find the length of the seconds pendulum on the moon, where $g = 150$ cm./sec^2.

5. A pendulum beating seconds at Greenwich, where $g = 32\cdot2$, is taken to a place where it loses 4 seconds per day. Find the value of g at this place.

C.   16

6. Iron expands ·0000066 of its length for every degree Fahrenheit rise of temperature. A clock with its pendulum bob suspended by an iron rod keeps correct time at 62° F. How many seconds per day will it lose in a temperature of 80° F.?

7. If a seconds pendulum be taken to the top of a mountain half a mile high, how many seconds will it lose in a day?

8. A seconds pendulum carried from sea-level to the top of a mountain loses 15 seconds a day. Taking the earth's radius at 4000 miles, find the height of the mountain.

9. Find the depth of a mine at the bottom of which a seconds pendulum loses 9 seconds a day.

CHAPTER XXIV.

224. THIS is the subject of Newton's immortal work, the *Principia*. We shall here extract the most fundamental propositions; but the student should on no account fail to consult the original, if only to turn over the pages and note the contents. In no other way is it possible to gain an impression of the immense range of applications made by Newton, and the almost superhuman power with which they are worked out. The propositions are given according to the numbering of the *Principia*, with slight simplifications from the original form.

225. PROPOSITION I. The equable description of areas.

When a body revolves in an orbit subject to the action of forces tending to a fixed point, the areas swept out by radii drawn to the fixed centre of force are in one fixed plane, and are proportional to the times of describing them.

Let the time be divided into equal parts, and in the first interval let the body describe the straight line AB with uniform velocity, being acted on by no force. In the second interval it would, if no force acted, proceed to c in AB produced, describing Bc equal to AB; so that the equal areas ASB, BSc, described by the radii AS, BS, cS drawn to the centre S, would be completed in equal intervals. But when the body arrives at B, let a centripetal force tending to S act upon it by a single instantaneous impulse, and cause the body to deviate from the direction Bc and to proceed in the direction BC.

Let cC be drawn parallel to BS, meeting BC in C; then at the

16—2

end of the second interval the body will be found at C (§ 153) in the same plane with the triangle ASB, in which Bc, cC are drawn.   Join SC, and the triangle SBC, between the parallels SB, Cc, will be equal to the triangle SBc, and therefore also to the triangle SAB.

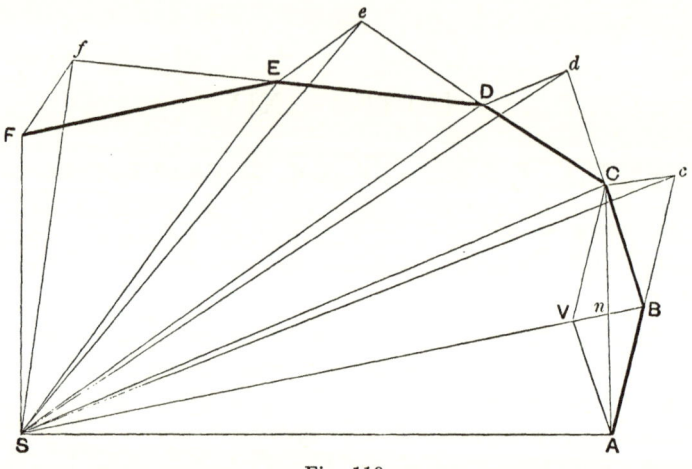

Fig. 119.

In like manner, if the centripetal force act upon the body successively at C, D, E, &c., causing the body to describe in successive intervals of time the straight lines CD, DE, EF, &c., these will all lie in the same plane; and the triangle SCD will be equal to the triangle SBC, and SDE to SCD, and SEF to SDE.

Therefore equal areas are described in the same plane in equal intervals; and the sums of any number of areas $SADS$, $SAFS$, are to each other as the times of describing them.

Let now the number of these triangles be increased and their breadth diminished indefinitely; then their perimeter ADF will be ultimately a curve line; and the instantaneous forces will become ultimately a centripetal force, by the action of which the body is continuously deflected from the tangent to this curve, and which will act continuously; and the areas $SADS$, $SAFS$, being always proportional to the times of describing them, will be so in this case.   Q. E. D.

The converse is given by Newton as his second proposition. It is easily seen that if the body moves so that the triangle SBC is equal to SBc, then cC must be parallel to BS, and therefore BS was the direction of the impulse at B. The property is then extended to the limiting case as before.

226. Kepler's Second Law, discovered from the motions of Mars, is ;—

The Case of the Planets.

The areas swept out by radii drawn from the planet to the sun's centre are, in the same orbit, proportional to the times of describing them.

Hence the planets move as if acted on by forces always directed to the sun's centre.

227. *Corollary.* Double the area swept out in the unit of time, one second, in any orbit is usually denoted by the letter h.

Since h is the same whether the body proceeds along the tangent with uniform velocity, when no force acts on it, or whether it is deflected round the curve, by a central force, it can be found if the velocity v is known. For let AB be the space described in one second, *i.e.*, v feet. Then $h = 2 \times$ area $SAB = v \times p$, where p is the perpendicular from S on the tangent at A.

Thus
$$v = \frac{h}{p},$$

and the velocity at any point is inversely proportional to the perpendicular from the centre of force on the tangent.

Given the velocity V, distance R, and direction of projection α, h can be found.

For $h = VR \sin \alpha$.

The time occupied in describing any part of the orbit is then obtained by dividing twice the area swept out by h.

Conversely, the position in the orbit after a time t has elapsed, is known from the area $\frac{1}{2}ht$ swept out in the interval.

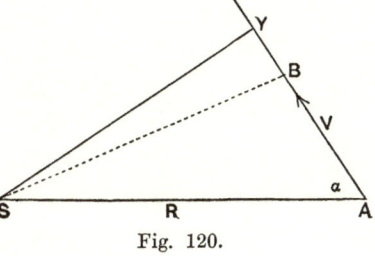

Fig. 120.

228. PROPOSITION VI. The Law of Variation of the Force.

Let the body move from P to Q in a very small interval of time t. Draw QR parallel and QT perpendicular to SP. Then (cf. § 111) if a is the acceleration towards S due to the force, supposed constant during the short interval, the distance QR fallen through towards the centre in the time t is

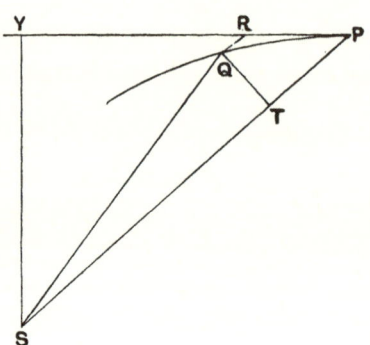

$$QR = \frac{at^2}{2}.$$

Fig. 121.

But
$$t = \frac{2 \text{ area } SPQ}{h}$$

$$= \frac{SP \cdot QT}{h}.$$

$$\therefore \; a = \frac{2QR}{t^2} = 2\,\frac{QR}{QT^2} \cdot \frac{h^2}{SP^2}$$

in the limit when PQ is indefinitely small.

The limiting value of $\dfrac{QR}{QT^2}$ depends on the shape of the curve, and the problem now is to find it for any desired case. Newton in his tenth and eleventh propositions finds it for (1) an ellipse described under a force tending to the centre, and (2) an ellipse described under a force tending to one focus.

229. PROPOSITION X. Let a body describe an ellipse; to find the law of force tending to the centre.

The acceleration $= \dfrac{2h^2}{CP^2} \cdot \dfrac{QR}{QT^2}$ (Prop. VI).

Draw the conjugate diameter CD, and the perpendicular PF.

By similar triangles QTu, PFC,

$$\frac{QT^2}{Qu^2} = \frac{PF^2}{CP^2}.$$

By the properties of the ellipse

$$\frac{Qu^2}{Pu.uG} = \frac{CD^2}{CP^2}.$$

$$\therefore \quad \frac{QT^2}{Pu.uG} = \frac{PF^2.CD^2}{CP^4} = \frac{AC^2.BC^2}{CP^4}.$$

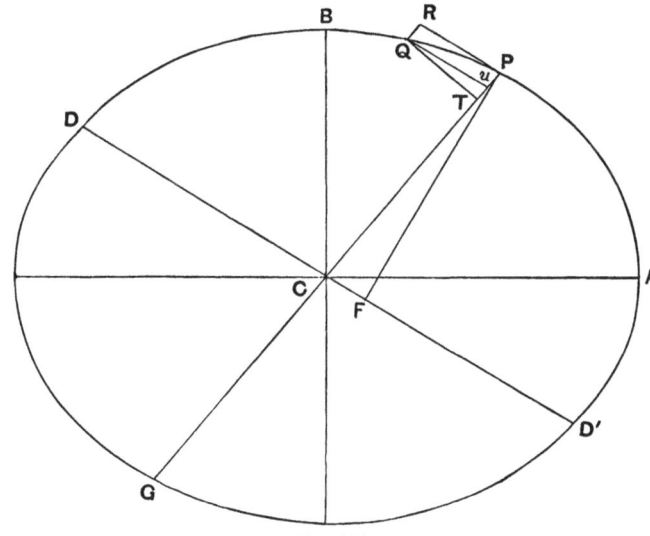

Fig. 122.

But $Pu = QR$, and ultimately, when Q is taken very close to P, $uG = 2CP$.

$$\therefore \quad \frac{QT^2}{2QR} = \frac{AC^2.BC^2}{CP^3} \quad \text{ultimately.}$$

$$\therefore \quad \text{the acceleration} = \text{limit of } \frac{2h^2}{CP^2} \cdot \frac{QR}{QT^2}$$

$$= \frac{h^2}{AC^2.BC^2} \cdot CP.$$

Denoting the semi-axes by a, b, we have

$$\text{acceleration} = \frac{h^2}{a^2b^2} \cdot CP.$$

The acceleration, and therefore the force producing it, is thus proportional to the distance CP, since $\dfrac{h^2}{a^2b^2}$ is a constant factor.

230. *Corollary* 1. Suppose the force acting on a body is such as to produce an acceleration μ times the distance of the body from a fixed point, and that the body has been projected with an initial velocity not directed towards the point. Then the body will describe an ellipse about that point as centre; and since

$$\text{acceleration} = \mu \cdot CP,$$

$$\frac{h^2}{a^2 b^2} = \mu,$$

$$\therefore \ h = \sqrt{\mu} \cdot ab.$$

The time of describing the complete ellipse

$$= \frac{2 \text{ area of ellipse}}{h} = \frac{2\pi ab}{\sqrt{\mu} \cdot ab} = \frac{2\pi}{\sqrt{\mu}}.$$

The periodic time is thus the same for all ellipses described about this centre of force, whatever their dimensions. The dimensions will depend on the distance and velocity of projection, but will not affect the periodic time.

Let different ellipses be described with their axes major along the same line; and let the bodies be given smaller and smaller velocities at right angles to this line. The ellipses will degenerate into lines along the axis, and the motions become vibrations about the centre with different amplitudes but the same periodic time. In fact we have arrived by a different path at the case of Simple Harmonic Motion, with its fundamental property that the period is independent of the amplitude (§ 215).

231. *Corollary* 2. The velocity v at any point

$$= \frac{h}{p} = \frac{h \cdot CD}{p \cdot CD} = \frac{h}{ab} \cdot CD = \sqrt{\mu} \cdot CD.$$

$$\therefore \ v = \sqrt{\mu} \cdot CD$$

and the velocity is proportional to the semi-conjugate diameter.

232. PROPOSITION XI. Let a body revolve in an ellipse; to find the law of force tending to a focus of the ellipse.

Let the force tend to the focus S; draw QR parallel and QT perpendicular to SP, and let Qxv, parallel to the tangent, cut SP in x and CP in v.

Draw HI parallel to CD.

Then $SE = EI$, since $SC = CH$;

and $PE = \frac{1}{2}(SP + PI) = \frac{1}{2}(SP + PH) = AC.$

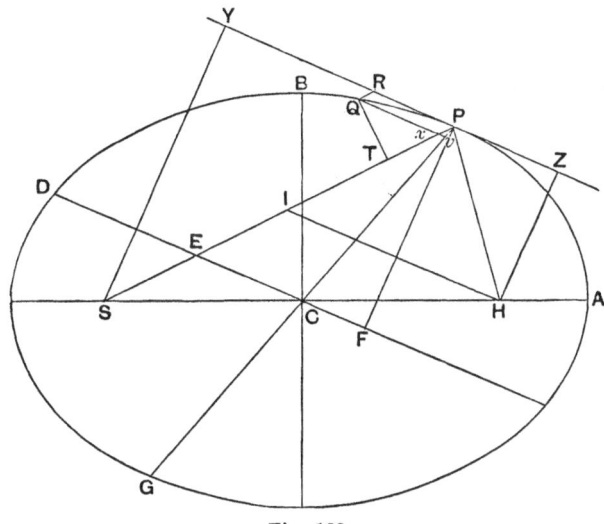

Fig. 123.

By Prop. VI. the acceleration = the limiting value of $\dfrac{2h^2}{SP^2} \cdot \dfrac{QR}{QT^2}$ when PQ is indefinitely diminished.

By similar triangles, QTx, PFE,

$$\frac{QT^2}{Qx^2} = \frac{PF^2}{PE^2} = \frac{PF^2}{AC^2} = \frac{BC^2}{CD^2} \quad \ldots\ldots\ldots\ldots\ldots(1)$$

since $PF . CD = AC . BC.$

By a property of the ellipse

$$\frac{Qv^2}{Pv . vG} = \frac{CD^2}{CP^2};$$

and by similar triangles

$$\frac{Pv}{QR} = \frac{Pv}{Px} = \frac{CP}{PE}.$$

$$\therefore \ Pv = \frac{CP}{AC} . QR,$$

and ultimately $\qquad\qquad vG = 2CP;\ Qx = Qv.$

$$\therefore\ Pv \cdot vG = \frac{2CP^2}{AC} \cdot QR,$$

and $\qquad\qquad \dfrac{Qx^2}{\dfrac{2CP^2}{AC} \cdot QR} = \dfrac{CD^2}{CP^2}$ ....(2).

Therefore multiplying (1) and (2),

$$\frac{QT^2}{\dfrac{2CP^2}{AC} \cdot QR} = \frac{BC^2}{CP^2}$$

and $\qquad\qquad \dfrac{QT^2}{QR} = \dfrac{2BC^2}{AC}.$

$\therefore$ the acceleration, which $= \dfrac{2h^2}{SP^2} \cdot \dfrac{QR}{QT^2},$

$$= \frac{h^2 \cdot AC}{BC^2} \cdot \frac{1}{SP^2}$$

$$= \frac{h^2 a}{b^2} \cdot \frac{1}{SP^2}.$$

The acceleration, and therefore the force producing it, is inversely proportional to the square of the distance SP from the focus, since $\dfrac{h^2 a}{b^2}$ is a constant factor.

233. Kepler's First Law is ;—

The planets move in ellipses having the sun in one focus.

The force exerted upon them by the sun must therefore be inversely proportional to the square of the distance.

This is the famous proposition which Hooke guessed at, but could not prove (§ 94).

234. *Corollary* 1. Suppose the force acting on a body P is such as to produce an acceleration $\dfrac{\mu}{SP^2}$ towards a point S. Then the body will describe an ellipse about S as a focus, such that

$$\frac{h^2 a}{b^2} = \mu.$$

$$\therefore\ h^2 = \mu \cdot \frac{b^2}{a}.$$

The time T of describing a complete ellipse

$$= \frac{2 \times \text{area of the ellipse}}{h}.$$

$$\therefore \ T = \frac{2\pi ab}{\sqrt{\mu} \cdot \frac{b}{\sqrt{a}}}$$

$$= \frac{2\pi}{\sqrt{\mu}} \cdot a^{\frac{3}{2}}.$$

Thus $T^2 \propto a^3.$

Kepler's Third Law is ;—

The squares of the periodic times are proportional to the cubes of the major axes.

It is to be inferred that the accelerations of all the different planets are such as would be experienced by any one of them at the same distance. In other words μ is the same for them all. If the earth were substituted for Jupiter, it would describe Jupiter's orbit.

Hence it is the same gravity which acts on all of them, a force proportional to their masses $\left(\text{since acceleration} = \dfrac{P}{M}\right)$ whatever may be the nature of their materials, and inversely proportional to the square of the distance from the sun.

235. *Corollary* 2. The velocity v is given by

$$v = \frac{h}{SY}.$$

Draw HZ the perpendicular from the other focus on the tangent. Then by the properties of the ellipse

$$\frac{SY}{SP} = \frac{HZ}{HP} = \sqrt{\frac{SY.HZ}{SP.HP}} = \frac{BC}{\sqrt{SP.HP}}.$$

$$\therefore \ SY = BC\sqrt{\frac{SP}{HP}}$$

and $v^2 = \dfrac{h^2}{SY^2} = \mu\,\dfrac{b^2}{a} \Big/ b^2\,\dfrac{SP}{HP}$

$$= \mu \cdot \frac{HP}{a.SP}.$$

Denote the distance SP by r.

Then
$$v^2 = \mu \cdot \frac{2a - r}{ar}$$

$$= \mu \left(\frac{2}{r} - \frac{1}{a} \right).$$

When the velocity of projection V, and the initial distance R from the centre of force, are known, the axis major is determined from

$$V^2 = \mu \left(\frac{2}{R} - \frac{1}{a} \right).$$

If $V^2 = \mu \cdot \dfrac{2}{R}$, a is infinite, and the ellipse becomes a parabola.

If $V^2 > \mu \cdot \dfrac{2}{R}$, a is negative, and the orbit is a hyperbola.

Newton points out these cases, but for the sake of their importance, repeats the proof of Prop. XI. in a form adapted to parabolic and hyperbolic orbits; and later on he shews that the orbits of comets are either hyperbolas (in which case they only once visit the solar system) or such extended ellipses that they approximate to parabolas, so that the comets only return at very long intervals.

236. The motions of all the heavenly bodies were thus accounted for on the hypothesis that they were acted on by central forces varying inversely as the squares of the distances and directed towards the sun.

Before this hypothesis could be generalized into the law of universal gravitation, Newton had to calculate how a sphere composed of particles each attracting every other particle with a force proportional to the product of the masses and inversely proportional to the squares of the distances would behave to another sphere similarly composed. This was effected in the beautiful series of propositions forming Section XII. of the *Principia*. The most important are here given with slight changes in accordance with modern methods. Newton had hardly hoped for such a simple result but had expected to treat the heavenly bodies as particles only as an approximation to the truth, permissible on account of their great distances.

237. Attraction of a thin spherical shell on a particle inside it. (*Principia*, Sec. XII. Prop. 70.)

Let the particle be at O.

Consider a cone of very small angle with O as vertex, cutting the surface at P and Q.

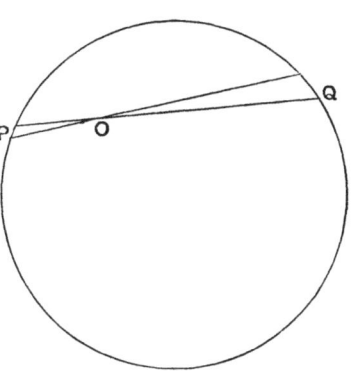

The surface is equally inclined to the chord PQ at P and Q, so that the areas cut out by the cone at P and Q are proportional to the squares of their distances from the vertex, *i.e.* as OP^2 is to OQ^2.

But the attractions of particles at P and Q are inversely as OP^2 and OQ^2. Therefore the attrac-

Fig. 124.

tions of the areas at P and Q are equal, for they are directly as the areas and inversely as the squares of the distances.

These attractions mutually destroy each other; and the same is true for every cone drawn through O.

Thus the attraction of the whole shell on O is zero.

And since a shell of definite thickness may be conceived as made up of a number of concentric thin shells, the same is true for the attraction of a thick shell upon a point inside its inner boundary.

238. Attraction of a thin spherical shell on a particle outside it. (*Principia*, Sec. XII. Prop. 71.)

Let P be the position of the particle, and O the centre of the sphere.

From symmetry the resultant attraction must be along PO.

Let ρ be the mass of the shell per unit area, r the radius of the sphere.

Join OP and divide it at B so that

$$OB : r :: r : OP.$$

Consider a cone through B cutting out a small area S at Q. Its attraction is

$$\frac{\rho S}{PQ^2} \text{ along } PQ.$$

The component of this along PO is

$$X = \frac{\rho S \cos OPQ}{PQ^2}.$$

But $\dfrac{OB}{OQ} = \dfrac{OQ}{OP}$; so that OBQ, OQP are similar triangles, and $\angle OQB = \angle OPQ$.

Also

$$\frac{BQ}{PQ} = \frac{OQ}{OP} = \frac{r}{OP}.$$

Thus

$$X = \frac{\rho r^2}{OP^2} \cdot \frac{S \cos OQB}{BQ^2}.$$

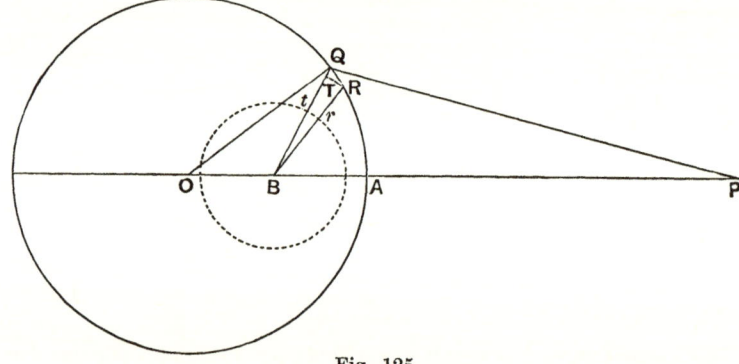

Fig. 125.

Let a sphere be drawn round B as centre, with radius BR, and another with any fixed radius a.

The cone will cut from these spheres the areas RT, rt.

But area $RT = S \cos QRT = S \cos OQB$.

$$\therefore\ X = \frac{\rho r^2}{OP^2} \times \frac{\text{area } RT}{BQ^2}$$

$$= \frac{\rho r^2}{OP^2} \times \frac{\text{area } rt}{a^2}.$$

Let the same be done for every small area of the shell. Then the total resultant attraction along PO

$$= \frac{\rho r^2}{OP^2} \times \frac{\text{whole surface of sphere, radius } a}{a^2}$$

$$= \frac{\rho r^2}{OP^2} \times \frac{4\pi a^2}{a^2}$$

$$= \frac{4\pi r^2 \rho}{OP^2}.$$

But this is the attraction of the mass of the shell $4\pi r^2 \rho$ collected at the centre O.

Hence the attraction of the shell is the same as if it were all collected at its centre.

Since a solid sphere may be conceived as made up of concentric shells, the same is true for a solid sphere, whether its density is the same throughout, or depends on the distance from the centre.

239. *Corollary.* (*Principia*, Prop. 73.)

The attraction of a homogeneous solid sphere upon a particle inside it is proportional to the distance from the centre.

For draw a concentric sphere through the point P distant r from the centre. Then the shell outside this sphere has no attraction on a particle at P (Prop. 70). The attraction of the rest is

$$= \frac{\frac{4}{3}\pi r^3 \rho}{r^2} \quad \ldots\ldots\ldots\ldots \text{(Prop. 71)}$$

$$= \tfrac{4}{3}\pi\rho \cdot r.$$

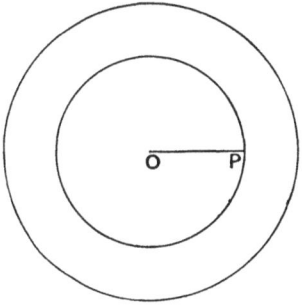

Fig. 126.

EXAMPLES.

1. If a straight tunnel could be driven through the centre of the earth, shew that a cannon ball dropped into it would reach the antipodes in about $42\frac{1}{2}$ minutes. Take $g = 32\cdot2$; earth's radius $= 4000$ miles ; and neglect the resistance of the atmosphere.

2. Shew that the time would be the same for a train of cars running on smooth rails through a straight tunnel to any part of the earth's surface.

3. Shew that if the earth's velocity in her orbit were increased by about one half, she would describe a parabola about the sun.

4. Shew that if a body were projected from the earth with a greater velocity than about 7 miles per second, it would not return to her.

5. The synodical period of a planet exterior to the earth, *i.e.* the interval between two successive conjunctions, or moments when they are in the same direction from the sun, is *S*. Shew that if *E* and *P* be the times of revolution of the earth and the planet about the sun,

$$\frac{1}{P} = \frac{1}{E} - \frac{1}{S}.$$

Hence find the time of revolution of Jupiter, whose synodical period is observed to be 398·88 days, the time for the earth being 365·25 days ; and from Kepler's Third Law deduce the mean distance of Jupiter from the sun, that of the earth being 92,390,000 miles.

CHAPTER XXV.

IMPACT AND IMPULSIVE FORCES.

240. ACCORDING to the Second Law of Motion the effect of a force is to determine at every instant a rate of change of velocity in a body on which it acts, *i.e.* an acceleration. From the acceleration we can find by the kinematic formulae, the velocity of the body at any subsequent time, and the distance it will have moved in the interval; and thus the whole effect of the force is known.

There are cases, as when two billiard balls collide, or a ball receives a blow from a cricket bat, where the whole time of action is so excessively short that we can no longer follow the process in detail; yet during the momentary contact forces are called into play, rising from zero to a high value, and dying away to zero again according to unknown laws, so that a great and apparently instantaneous change of velocity takes place. Such forces are called impulsive, and these cases of impact or collision require a somewhat different treatment.

The very fact that prevents us from applying the previous method, the shortness of the time, relieves us of half the difficulty. It is not necessary to calculate where the body will be, since it has not time to change its place appreciably during the blow. We know where it is, and it suffices to find the total change of velocity. The subsequent motion under finite forces can then be calculated as before.

241. To fix our ideas, let us suppose that a sphere A of mass M, moving to the right with velocity U overtakes and impinges directly upon another sphere B of mass m, also moving to the right with smaller velocity u. By direct impact, is meant that the line of centres is the line of motion for both balls.

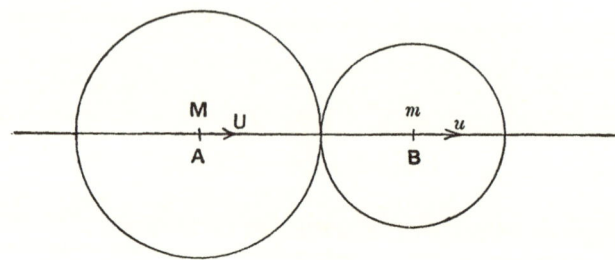

Fig. 127.

The blow will change both velocities. Let the new velocities, immediately after impact, be V and v.

However irregularly the pressure between the balls may vary during the impact, at every instant the action of the first ball upon the second will, by the Third Law of Motion, be met by an equal and opposite reaction of the second upon the first. And therefore the total impulse for the whole time of contact will be the same for the reaction as for the action, but in the opposite direction.

By the Second Law change of momentum is equal to the impulse; so that the momentum of B will be increased precisely as much as the momentum of A is diminished, and the sum of the momenta after impact is the same as it was before. Thus

$$MV + mv = MU + mu.$$

Or instead of considering the effect on each ball separately, let the system considered be the two balls taken together. During the infinitely short time of contact any finite external forces that may be acting, such as gravity, have no time to produce a change of momentum. The momentum of the system is therefore unaltered, and again

$$MV + mv = MU + mu \dots\dots\dots\dots\dots\dots\dots(1).$$

A second relation is required to determine the two unknown quantities V and v. For this we recur to experiment.

242. If the balls are made of clay, putty, or similar substances,

I. Inelastic Bodies.   they are squeezed out of shape by the blow, but shew no tendency to recover their form and to thrust each other apart. They adhere together and move forward as one mass.

In this case

$$V = v \dots\dots\dots\dots\dots\dots\dots\dots(2).$$

Solving (1) and (2) we have

$$V = v = \frac{MU + mu}{M + m} \cdot$$

The total impulse R between the balls is equal to the change of momentum of either ball. Thus

$$R = m(v - u) = -M(V - U)$$
$$= \frac{Mm(U - u)}{M + m} \cdot$$

243. Balls of ivory, steel, glass, and most other substances instantly recover their form and thrust each other apart. Newton discovered the second relation for these elastic substances by means of an experiment remarkable for its simplicity and elegance.

II. Elastic Bodies.

Experiment.

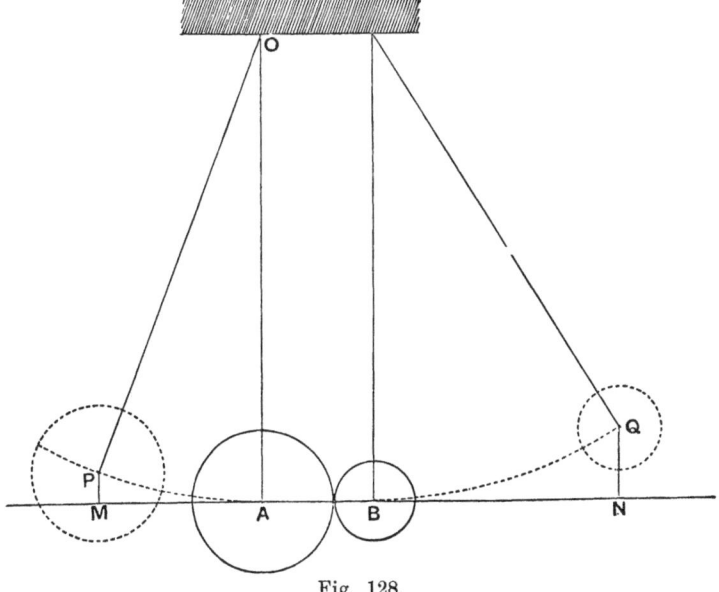

Fig. 128.

17—2

Let two balls be suspended from a lofty support by long threads in such a manner that they rest in contact with their centres in the same horizontal line when the threads are precisely equal and parallel. To ensure that they shall move in the same plane it is best to use a **V**-shaped suspension instead of a single thread for each ball.

Draw the balls aside through different arcs, and release them simultaneously by gently opening the fingers. They will meet at the lowest point. For by the property of the pendulum (§ 216) the time of describing any arc, small compared with the length of the suspension, is independent of the length of the arc.

The velocities with which they meet are easily calculated. For they are those due to the vertical falls. Thus

$$v^2 = 2g \cdot PM.$$

But $PM \cdot MM' = MA^2$, and since AP is small compared with the radius $OA \ (=a)$,

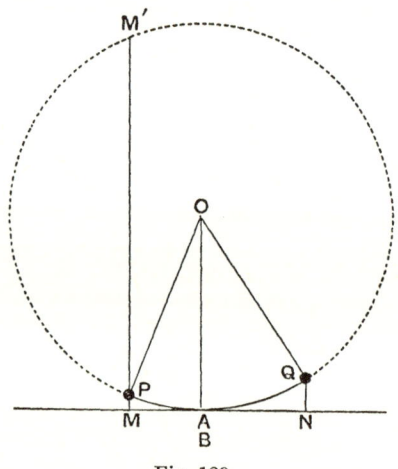

$$PM = \frac{MA^2}{2a} \text{ very approximately.}$$

$$\therefore \quad v^2 = \frac{2g}{2a} \cdot AM^2 = \frac{g}{a} \cdot AM^2,$$

and v is proportional to AM.

Set a metre scale horizontally just beneath the two balls, so that when they are drawn aside to P and Q, the

Fig. 129.

distances AM, BN can be read off; let them be H, h. These will serve as measures of the velocities at impact.

Similarly, let the horizontal distances K, k, to which the balls rebound, be noted. These measure on the same scale, the velocities of rebound at A, B, since a ball will rise on a curve to that height from which it must fall to gain its velocity of projection.

With a little practice two observers, one to release the balls simultaneously from measured distances AM, BN, and observe the distance K of rebound, while the other observes the rebound k, can obtain very consistent results for a series of experiments, and still greater accuracy may be attained if the balls are held in position by a very fine thread, and released from rest by burning it, instead of releasing them by hand.

Let H, h, K, k be given the signs of the velocities they measure. Thus H and k representing velocities to the right will be positive, and h and K negative. Then it will be found that

$$(K - k) = - e (H - h),$$

or, *numerically*, $(K + k) = e (H + h)$, where e is a constant fraction for all values of H and K, so long as balls of the same material are used.

Translating this into velocities, we have

$$(V - v) = - e (U - u),$$

i.e., the relative velocity after impact is a fixed fraction of the relative velocity before impact, and is reversed in direction.

If balls of other materials are used, the value of the fraction e will be different, but the same law will hold good. e is called the *coefficient of restitution* for the given materials.

This is the relation discovered by Newton. The beautiful ingenuity with which the property of the pendulum, and Galileo's theory of motion on a smooth curve are employed, will be best appreciated by the student if he will try to think how else he could project two balls so as to be sure they will meet at an expected point where they may be observed; control and vary the velocities of impact; and measure the instantaneous velocities of rebound.

244. When the coefficient e has been experimentally determined for balls of given materials, the problem of collision is easily solved.

By the dynamical principle of momenta we have

$$MV + mv = MU + mu\dots\dots\dots\dots\dots\dots(1).$$

By Newton's experimental result

$$V - v = - e (U - u)\dots\dots\dots\dots\dots\dots(2).$$

Whence $$V = \frac{MU + mu - em(U - u)}{M + m},$$

$$v = \frac{MU + mu + eM(U - u)}{M + m}.$$

For inelastic bodies $e = 0$, and as in § 242

$$V = v = \frac{MU + mu}{M + m}.$$

For balls of glass $e = 0{\cdot}94$.
   „   „   „ ivory $e = 0{\cdot}81$.
   „   „   „ cast iron $e = 0{\cdot}66$.
   „   „   „ lead $e = 0{\cdot}2$.

245. The physical meaning of the coefficient e may be seen as follows.

Before impact the centres of gravity of the balls are approaching, and after it they are separating. There must have been some moment during impact at which they were relatively at rest, and the balls on the whole had the same velocity. Let us call this the moment of greatest compression. Let R be the total impulse between the balls up to that moment, *i.e.* during compression; and R' the further impulse while they are recovering their shape, *i.e.*, during restitution.

At the moment of greatest compression both balls have the same velocity V'; hence, as for inelastic bodies,

$$(M + m)\, V' = MU = mu,$$

$$V' = \frac{MU + mu}{M + m}.$$

The impulse R is measured by the change of momentum of either ball up to the moment of greatest compression. Taking the first ball

$$R = MV' - MU$$

$$= M \cdot \frac{MU + mu}{M + m} - MU$$

$$= -\frac{Mm(U - u)}{M + m}.$$

The impulse R' is the further change of momentum after the moment of greatest compression.

$$R' = MV - MV'$$
$$= M\frac{MU + mu - em(U-u)}{M+m} - \frac{MU+mu}{M+m}$$
$$= -\frac{eMm(U-u)}{M+m}$$
$$= -eR.$$

The fraction e thus measures the ratio of the impulse of the elastic forces by which the balls recover their form to the impulse of the force used in compressing them.

246. The energy before impact is

Loss of Energy during an impact.

$$E = \tfrac{1}{2}(MU^2 + mu^2).$$

The energy after impact is

$$E' = \tfrac{1}{2}(MV^2 + mv^2)$$
$$= \frac{M}{2}\left\{\frac{MU+mu-em(U-u)}{M+m}\right\}^2 + \frac{m}{2}\left\{\frac{MU+mu+eM(U-u)}{M+m}\right\}^2$$
$$= \frac{(M+m)(MU+mu)^2 + (m+M)e^2 . Mm(U-u)^2}{2(M+m)^2}$$
$$= \frac{(MU+mu)^2 + e^2 Mm(U-u)^2}{2(M+m)}.$$

If $e=1$, i.e. if the balls are perfectly elastic,

$$E' = \frac{(MU+mu)^2 + Mm(U-u)^2}{2(M+m)}$$
$$= \frac{M^2U^2 + m^2u^2 + MmU^2 + Mmu^2}{2(M+m)}$$
$$= \frac{MU^2 + mu^2}{2}$$
$$= E.$$

Thus the total kinetic energy of the two balls is the same after impact as before.

In practice e is always less than 1. Then

$$E' = \frac{(MU+mu)^2 + Mm(U-u)^2}{2(M+m)} - \frac{(1-e^2)Mm(U-u)^2}{2(M+m)}$$
$$= E - \frac{(1-e^2)Mm(U-u)^2}{2(M+m)}.$$

The second term on the right-hand side is necessarily positive, so that E' is less than E.

The energy of motion after impact is thus less than it was before. The rest has been transformed into energy of heat and sound, or permanent deformation of the body against its cohesive forces.

247. Whether the energy so transformed is to be considered as lost or not, depends upon the purpose with which the blow is struck. If it be desired to drive a pile into the ground, or a nail into a block of wood, the energy converted into sound, heat, and permanent deformation is wasted. But for shaping a rivet by hammering, or forging a block under the steam-hammer, this contains the valuable part of the energy.

In practice the object struck is generally at rest, and may be taken as inelastic. In this case $u = 0$; $e = 0$; and the second term in the expression for E' becomes

$$\frac{MmU^2}{2(M+m)}.$$

The energy of the hammer was $\dfrac{MU^2}{2}$.

Thus $\qquad \dfrac{\text{Energy transformed}}{\text{Total energy}} = \dfrac{m}{M+m}.$

For driving piles or nails this must be made as small as possible; so that M should be great compared with m; i.e., the ram of the pile-driver or the hammer-head, should have great mass compared with the pile or nail to be driven.

For shaping rivets and forgings $\dfrac{m}{M+m}$ must be as large as possible; i.e., m must be large compared with M. This means that the anvil must be much heavier than the hammer. The hammer again should be heavy with slow velocity, rather than light with high velocity, in order to avoid waste of energy in sound and heat, and convert as much as possible into permanent deformation.

These were the considerations which guided Nasmyth in the construction of his steam hammer. He says * :—

* Wright's *Mechanics.*

" Pile driving had before been conducted on the cannon ball principle. A small mass of iron was drawn slowly up, and suddenly let down on the head of the pile at a high velocity. This was destructive, not impulsive action. Sometimes the pile was shivered into splinters without driving it into the soil ; in many cases the head of the pile was shattered into matches, and this in spite of the hoop of iron about it. On the contrary I employed great mass and moderate velocity. The fall of the steam hammer block was only 3 or 4 feet, but it went on at 80 blows a minute, and the soil into which the pile was driven never had time to grip or thrust it up."

248.   In this case v and u are both zero in the equation

$$V - v = - e\,(U - u),$$

Impact of a sphere on a fixed plane.   so that $$V = - eU,$$

and the sphere is reflected with velocity diminished in the ratio $e : 1$.

The impulsive pressure between the sphere and the plane is measured by the change of momentum produced in the sphere, and is equal to

$$MV - (-eMV)$$
$$= MV\,(1 + e).$$

For example, let a stream of water one inch in diameter strike a wall directly with velocity v feet per second.

The volume of water striking the wall per second is $v \times \pi \left(\frac{1}{24}\right)^2$ cubic feet, and its mass at 1000 oz. per cubic foot is

$$v\pi \left(\tfrac{1}{24}\right)^2 . \tfrac{1000}{16} \text{ lbs.}$$

The change of momentum per second is

$$mv\,(1 + e) = \tfrac{1000}{16}\,\pi \left(\tfrac{1}{24}\right)^2 . v^2\,(1 + e),$$

and this measures the steady pressure between the water and the wall.

In hydraulic mining the impact of powerful streams of water is largely employed for cutting away hill sides.

Very important applications of this theory are found in the Kinetic Theory of Gases, in which the known relations between the volume, pressure, and temperature of a gas are explained by regarding it as an assemblage of innumerable free particles in rapid motion.

249. Cases of oblique impact between smooth bodies are treated by resolving the motions along the line

Oblique Impact.   of centres and perpendicular to it. The velocities along the line of centres can be calculated by the laws of direct impact; the velocities at right angles to the line of centres are unchanged since there is no friction. The resultant final velocities are then found by compounding.

The case of impact on a plane will suffice for illustration.

Let u be the velocity with which a particle moving along PO strikes a fixed plane at O. This is equivalent to $u \cos \alpha$ perpendicular to the plane, and $u \sin \alpha$ parallel to it.

After impact the velocities will be $-eu \cos \alpha$, and $u \sin \alpha$ respectively. The resultant velocity will be

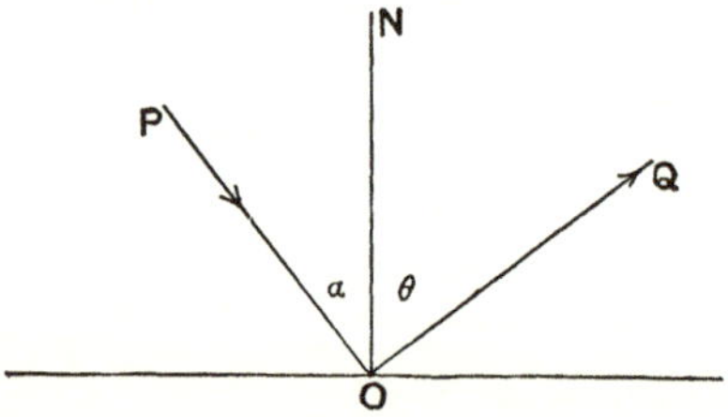

Fig. 130.

$$u \sqrt{e^2 \cos^2 \alpha + \sin^2 \alpha},$$

and the direction of motion will make an angle θ with the normal, where

$$\tan \theta = \frac{u \sin \alpha}{e \cdot u \cos \alpha} = \frac{1}{e} \tan \alpha.$$

EXAMPLES.

1. A mass of 10 lbs. moving 5 feet per second overtakes a mass of 4 lbs. moving in the same line 3 feet per second. Find the velocities after impact, and the total impulse between the masses (1) when they are inelastic, (2) when the coefficient of restitution is ·6.

2. Two balls ($e = \frac{2}{3}$) of masses 4 lbs. and 3 lbs. impinge from opposite directions with velocities of 6 and 8 feet per second respectively. Find the velocities after impact, and the loss of kinetic energy.

3. Shew that if two perfectly elastic balls of equal mass impinge directly they exchange velocities.

4. A perfectly elastic ball is projected against the first of a number of exactly similar balls arranged in a straight line, each in contact with the next. Shew that the last ball will fly off with the velocity of the impinging ball, the others remaining at rest.

5. A train of cars loaded to equal weights is standing at rest with a space of three inches between each car and the next, the utmost the couplings will allow. Another car of equal weight is shunted on to it behind at 1 mile an hour. Shew that if the ·buffers are perfectly elastic and the couplings inelastic, a passenger will experience two forward jerks before getting into uniform motion, and that, if each car is 60 feet long, the mean speed with which the first impulse travels through the train is 241 miles an hour. Find also the interval between the two jerks for a passenger in the sixth car from the front ; and the velocity with which the train finally starts off if there are 20 cars in all.

6. A fire engine can project a stream of water $1\frac{1}{2}$ inches in diameter to a height of 144 feet. If the stream is turned horizontally on to a wall find the pressure on the wall, taking $e = \frac{1}{2}$, and a cubic foot of water to weigh 1000 ounces.

7. The ram of a pile-driver weighs 200 lbs. and drives a pile $\frac{3}{4}$ inch into the ground after falling 16 feet. What steady pressure could the pile support ?

8. How many blows of a steam hammer weighing 400 lbs. (with a stroke of 2 ft., pressure of steam 80 lbs. per square inch, and piston diameter 8 inches) would be required to drive the pile in question 7 another 6 inches into the ground ?

9. Shew that to hit a billiard ball after one reflection from a cushion you should aim at an imaginary ball as far behind the cushion as the real one is in front of it, assuming that the cushions are perfectly elastic.

10. In the "half-ball" stroke at billiards the player aims so that the centre of his ball would pass through the extreme edge of the ball aimed at. Assuming the balls perfectly elastic, find the inclinations of their directions of motion after impact to the line joining their centres before either was struck, if the distance between centres was 5 feet and the diameter of the balls 2 inches.

11. A projectile weighing 600 lbs. is fired from a gun weighing 12 tons with a muzzle velocity of 2000 feet per second. What is the velocity of recoil of the gun ?

12. If the earth, when at the end of the minor axis of her orbit, collided directly with a comet of one-millionth of her mass and twice her velocity, and absorbed it, what would be the change in the major axis of her orbit ? Hence find the change in the length of the year.

BOOK IV.

THE ELEMENTS OF RIGID DYNAMICS.

CHAPTER XXVI.

THE COMPOUND PENDULUM.

250 Up to this point the masses whose motions have been investigated have been supposed to be collected at single points, capable only of a motion of translation from place to place. This part of the subject is called the Dynamics of a Particle.

What is known as Rigid Dynamics, or the Dynamics of a Rigid Body, took its rise out of the problem of the Compound Pendulum. Neither particles (in the sense of masses collected at mathematical points) nor absolutely rigid bodies are found in nature. But, so far as motion of translation is concerned, real bodies behave as if their masses were collected at their centres of gravity; and most solids are sufficiently rigid to justify their treatment as perfectly rigid bodies, at all events for a first approximation. We can afterwards go on to take account of their deformations, and then we enter on the Theory of Elasticity.

The Simple Pendulum treated in Chapter XXIII consisted of a bob, whose mass was supposed to be collected at its centre, suspended by a string without weight. We cannot construct such an ideal pendulum, but we can go very near it by taking a very heavy bob, and suspending it by a fine but strong wire.

Any solid object suspended from a horizontal axis is found to perform oscillations exactly like those of a simple pendulum.

Drill a hole through a flat board, and pass a knitting needle through it. Hang a bullet just in front of the board by a thread wound round the needle. Hold the needle horizontal, and set them swinging by a jerk to one side. If the time of swing of

the bullet is less or greater than that of the board, unwind or
wind up the thread till they become
equal. Mark the spot on the board
exactly behind the bullet when both
are at rest. Then set them swinging
by another jerk. It will be found
that the bullet remains on the marked
spot throughout the motion; nor can
it be made to leave the spot by any
amount of jerking, however irregular.
As dynamical systems oscillating
about this axis under gravity the
board and the suspended bullet are
identical.

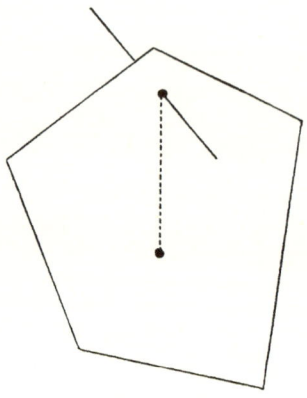

Fig. 131.

Such a solid object as the board
is called a *Compound Pendulum*. An ideal pendulum of the same
time of swing as the bullet is called the *Simple Equivalent Pendulum* for the board.

251. The problem, first solved by Huyghens, but attempted
by most of the leading mathematicians of his time, was to find by
calculation the time of swing of a compound pendulum of any
shape. The difficulty to be overcome was this. The particles of
the object at different distances from the axis would, if free to
swing separately, perform oscillations in different times, those near
the axis vibrating rapidly, and those far away more slowly. But
they are constrained, as parts of a rigid body held firmly together
by internal forces, all to vibrate in the same time. A compromise
has to be effected, and the whole body swings at some intermediate
rate, which in fact may be determined by the experiment with the
bullet. How can this rate, or the length of the simple equivalent
pendulum, be calculated from the known dimensions and structure
of the body, and the position of the axis?

The idea by which Huyghens reached his solution was this.
Suppose that at some moment when the body is passing through
its lowest position, *i.e.*, when the centre of gravity is in the vertical
through the axis, the whole body could be released from the
internal forces, and resolved into its separate particles, so that

each could swing on its own account. Then the height through which the centre of gravity of the system of separate pendulums formed by the free particles will rise in consequence of their existing motions must be the same as the height to which the centre of gravity of the body as a whole actually rises. Huyghens sees this as an extension of Galileo's principle of Work (§§ 41, 66). There cannot be a rise of weights on the whole, effected on their own account, without the aid of external forces. If the centre of gravity rose more, or less, in the one case than in the other, it would be possible to make a perpetual motion and even produce work out of nothing, as Galileo argued in the case of motion on an Inclined Plane (§ 66).

252. For simplicity, consider a straight rod free to swing
Problem of the   about one end C. Let the rod swing from CB
Straight Rod.   to CB', and let $CO = L$ be the length of the simple
equivalent pendulum, so that a bullet hung by a thread CO would swing to CO'.

Let v be the velocity of the bullet as it passes through O. At this moment the velocity of any other point in the rod, distant r from the axis C, is

$$\frac{r}{L} \cdot v.$$

Suppose that the rod is now broken up into its separate particles, each being left free to swing as a simple pendulum about C. The vertical height through which O, or the bullet, will rise is $\frac{v^2}{2g}$ (§ 111). The particle at distance r will rise through a height $\frac{r^2}{L^2}\frac{v^2}{2g}$.

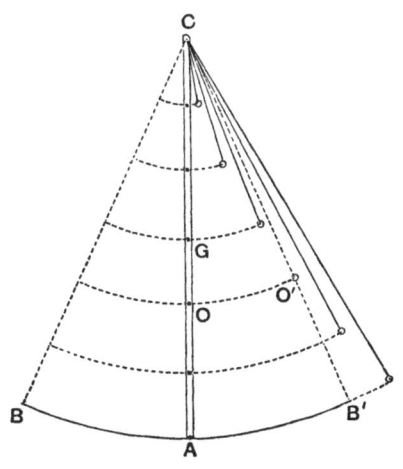

Fig. 132.

c.   18

Let m be the mass of this particle. Then the rise of the centre of gravity of the whole system of free particles will be

$$\frac{\Sigma m \cdot \dfrac{r^2 \cdot v^2}{L^2 \cdot 2g}}{\Sigma m} \quad (\S\S\ 21,\ 43) \dots\dots\dots\dots (1).$$

Let G be the centre of gravity of the rod, distant $\bar{x}$ from C. In the actual oscillation this rises through a vertical height

$$= \frac{\bar{x}}{L} \times \text{rise of } O$$

$$= \frac{\bar{x}}{L} \cdot \frac{v^2}{2g} \quad\dots\dots\dots\dots\dots\dots\dots\dots\dots\dots\dots (2).$$

By Huyghens' principle (1) and (2) must be the same. And since $\bar{x} = \dfrac{\Sigma m r}{\Sigma m}$ ($\S$ 21), we have

$$\frac{\Sigma m \cdot \dfrac{r^2}{L^2} \cdot \dfrac{v^2}{2g}}{\Sigma m} = \frac{\bar{x}}{L} \cdot \frac{v^2}{2g} = \frac{\Sigma m r}{L \cdot \Sigma m} \cdot \frac{v^2}{2g},$$

and, clearing out the constant factors $\dfrac{v^2}{2g}$, Σm, L,

$$\frac{\Sigma m r^2}{L} = \Sigma m r,$$

so that

$$L = \frac{\Sigma m r^2}{\Sigma m r}.$$

By the law of the Simple Pendulum, the time of oscillation of the rod is therefore

$$2\pi \sqrt{\frac{L}{g}} = 2\pi \sqrt{\frac{\Sigma m r^2}{g \cdot \Sigma m r}}$$

$$= 2\pi \sqrt{\frac{\Sigma m r^2}{Mg \cdot \bar{x}}},$$

where $M = \Sigma m$, the total mass of the rod.

The problem is thus reduced to the calculation of the quantity $\Sigma m r^2$ for the rod. Nowadays this is easily effected for a body of any regular shape by the Integral Calculus. But in default of modern analysis Huyghens employed the following ingenious device.

253. Let CA be the vertical rod. Draw $AD = AC$, and imagine a triangular flat plate CAD of such thickness that the mass per unit length of any thin slice PQ parallel to the base is equal to the mass m of the small portion of the rod opposite to it. Then if $CP = PQ = r$, the mass of PQ is mr, and its distance below C is r.

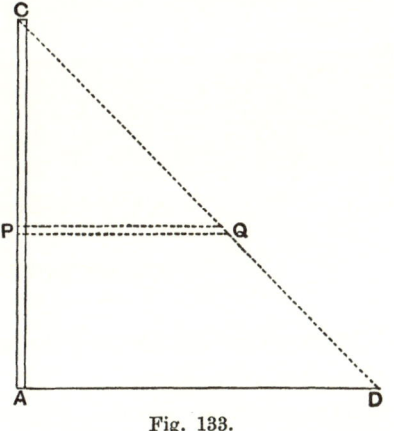

Fig. 133.

The c.g. of the triangular plate is at a distance below C

$$= \frac{\Sigma mr \cdot r}{\Sigma mr} = \frac{\Sigma mr^2}{\Sigma mr}.$$

But this is the value of L for a uniform rod. The c.g. of the triangle is known to be at a vertical distance $\frac{2}{3}CA$ below C.

Thus $\qquad\qquad L = \frac{2}{3} \cdot CA.$

EXAMPLES.

1. A uniform rod suspended by one end makes 75 oscillations per minute. Find the distance from this end of another point about which it would also make 75 oscillations per minute.

2. Find the length of a uniform rod which would beat seconds when suspended by one end.

CHAPTER XXVII.

D'ALEMBERT'S PRINCIPLE.

254. THE difficulties which Huyghens surmounted in his *Horologium Oscillatorium*, 1673, for the special problem of the Compound Pendulum, arise with ever increasing complication in other cases of the motion of rigid bodies.

Seventy years afterwards D'Alembert (*Traité de Dynamique*, 1743) gave a general principle by which all such problems may be treated. It was, perhaps, not so much a new principle, as an ingenious device for reducing the problems of Rigid Dynamics to the familiar laws already established for the equilibrium of forces.

After all, a rigid body consists ultimately of separate particles; and the motion of each of these is determined according to the Second Law of Motion by the resultant of the forces acting on it, *including the reactions from its neighbours* which hold it in position. The difficulty consists in the number of particles to be considered, and the unknown nature of these internal reactions between them. D'Alembert shewed how to avoid the consideration of separate particles and internal reactions.

Consider a single particle of mass m. It will be subject to (1) its weight and other gravitational, electric, or magnetic attractions and repulsions, and perhaps some tensions and pressures directly applied to it from outside the body. Call these the *impressed*, or *external*, forces. (2) The forces which bind it to its neighbours, and hold it in position as part of the rigid body. Call these the *internal* forces.

Since all the forces, external and internal, are applied to a particle, they must have a resultant. Let this be P.

By the Second Law of Motion the acceleration a of the particle will be in the direction of this resultant and equal to $\dfrac{P}{m}$; so that $ma = P$.

If, then, we could apply to the particle an extra force ma, *reversed*, *i.e.* in the opposite direction to P, the whole set of forces acting on the particle, including the reversed ma, would be in equilibrium.

Let every particle in the body be treated in the same way, *i.e.* let a force be applied to it equal to its mass multiplied by its acceleration, reversed in direction.

Then since the forces acting on each separate particle will be in equilibrium, so also will the whole system of forces acting on the body. This system consists of three groups:

(1) The set of reversed forces of the type ma. Indicate these by Σma.

(2) The set of internal forces.

(3) The impressed or external forces wherever applied to the body.

Now D'Alembert points out that the second group consists of pairs of equal and opposite forces. For if a particle A exerts an action R on a neighbouring particle B, then by the Third Law of Motion B exerts an equal and opposite reaction R on A. When, therefore, the internal forces are summed for the whole body, *they must form a system in equilibrium by themselves*, and exactly neutralize each other.

It follows that the first group must balance the third.

Hence the system of the reversed forces Σma is in equilibrium with the forces impressed on the body from outside.

255. This is D'Alembert's Principle. It is applied as follows. Convenient expressions may be found for Σma in terms of (1) the acceleration of the centre of gravity of the body, and (2) the angular accelerations of the body about its centre of gravity. Employing these expressions we write down the conditions of equilibrium for Σma and the impressed forces according to the rules of Statics. The resulting equations are sufficient to determine the motion of the centre of gravity, and the rotation of the body about the centre of gravity; and thus its motion is completely determined.

The problems of Rigid Dynamics are often extremely difficult on account of the complexities of Solid Geometry and Calculus required for their solution. Fortunately some of the most important are also the simplest. We shall give here one or two illustrations of D'Alembert's Principle, beginning with the solution of Huyghens' problem of the Compound Pendulum, for the sake of some experimental results. For modern developments of the subject the student must consult treatises on Rigid Dynamics.

256. Let a body of any shape swing about a horizontal axis

The Compound Pendulum by D'Alembert's Principle.

through C. Let θ be the inclination to the vertical of a line through C and the centre of gravity G.

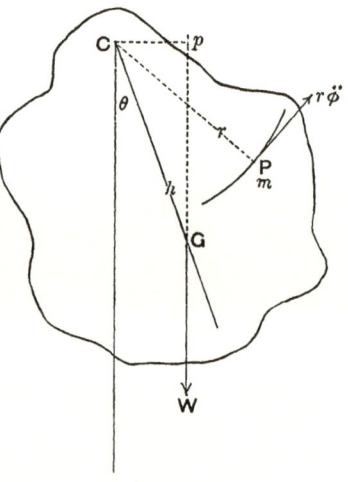

Fig. 134.

The angular velocity of CG about C is the rate of increase of θ per unit of time. If it is not constant, we proceed exactly as for variable velocities in a straight line (§ 71), and take the ratio of the very small angle $d\theta$ (or difference of θ) to the small increment of time in which it is described. A variable angular velocity is thus measured by the limiting value of

the fraction $\dfrac{d\theta}{dt}$, when the time dt is made indefinitely small. Denote this by $\dot{\theta}$.

As in the case of rectilinear velocities, we proceed to measure angular accelerations by the change in value of the angular velocity per unit of time, and in order to include the case of variable acceleration, choose an indefinitely small time in which the velocity alters.

Angular acceleration is thus measured by the limiting value

of $\dfrac{d\dot\theta}{dt}$ when dt is taken indefinitely small. Call this $\ddot\theta$. In the language of the Differential Calculus it is written $\dfrac{d^2\theta}{dt^2}$.

Let m be the mass of the particle at P distant r from the axis. Then if angles be measured in circular measure, the arc described by P while CP moves through an angle ϕ is $r\phi$; the linear velocity of P along the arc is $r \times$ angular velocity of $CP = r\dot\phi$; and the acceleration of P along the arc is $r\ddot\phi$.

But all lines from C to particles in the body must have the same angular motion about C. Thus $\ddot\phi = \ddot\theta$ for every particle.

We can now express D'Alembert's Principle. We are to apply to every particle m a force equal to its mass $\times$ its acceleration reversed, and then express the condition that all these forces balance the external forces. As this is a case of rotation about an axis or fulcrum, the latter condition must be that the sum of the moments of all the forces about the axis is zero.

Besides the acceleration $r\ddot\phi$ along the tangent, the particle has an acceleration towards C $(= r\dot\phi^2, \S\,77)$. But this we need not consider, since it will have no moment about C. The moment of the force $mr\ddot\theta$ about C is $mr\ddot\theta \times r$. The moment of all such forces, reversed in direction, for every particle of the body will be $-\Sigma mr\ddot\theta \,.\, r = -\ddot\theta\Sigma mr^2$, since $\ddot\theta$ is the same for all of them.

External Forces. The external forces are (1) the reaction at the axis C, and (2) the weights of all the particles.

The former has no moment about the axis.

The latter are equivalent to the whole weight of the body supposed collected at G. Let the distance CG, from the axis to the centre of gravity, be h, and let M be the mass of the body. Then its weight is Mg, and its moment about the axis is $Mg \,.\, h \sin\theta$. This is to be counted negative since it is in the clockwise direction.

By D'Alembert's Principle

$$-\ddot\theta\Sigma mr^2 - Mgh \sin\theta = 0,$$

$$\therefore\ \ddot\theta = -\frac{Mgh}{\Sigma mr^2} \sin\theta.$$

As in the case of the Simple Pendulum, let us suppose that

θ is a small angle such that its circular measure may be put for its sine.   Then

$$\ddot{\theta} = - \frac{Mgh}{\Sigma mr^2} \cdot \theta,$$

or, the angular acceleration is proportional to the angular displacement from the position of rest, and tends towards it.

But this is the law of Simple Harmonic Motion (§ 208).

The pendulum will therefore describe a Simple Harmonic Vibration of periodic time

$$2\pi \sqrt{\frac{\Sigma mr^2}{Mgh}}.$$

This is Huyghens' result.

EXAMPLES.

1.   A compound pendulum with mass M, moment of inertia (§ 257) about the axis I, distance of centre of gravity from axis h, is released when the line through the axis and the centre of gravity is inclined a to the vertical.   Shew that the angular velocity ω when this line is inclined θ to the vertical is given by

$$\omega^2 = \frac{2Mgh}{I} (\cos \theta - \cos a).$$

(Equate the energy (§ 260) of the pendulum to the work done.)

2.   Shew that the angular velocity of the simple equivalent pendulum of length L, released simultaneously at the same inclination a, is given by

$$\omega^2 = \frac{2g}{L} (\cos \theta - \cos a).$$

3.   Deduce the formula for the length of the simple equivalent pendulum by comparing the results of questions 1 and 2.

CHAPTER XXVIII.

MOMENT OF INERTIA.

257. THE quantity Σmr^2, which occurs in all problems connected with rotation about an axis, was called by Euler the Moment of Inertia of the body about the axis. It is the sum of the products of each element of mass by the square of its distance from the axis. The following Table gives its value for the principal regular figures as calculated by the Integral Calculus. We shall see later how Moments of Inertia may be determined experimentally.

Table of Moments of Inertia.

The moment of inertia of

(1) A rectangle whose sides are $2a$, $2b$

about an axis through its centre in its plane perpendicular to the side $2a$ $= \text{mass} \times \dfrac{a^2}{3}$,

about an axis through its centre perpendicular to its plane $= \text{mass} \times \dfrac{a^2 + b^2}{3}$.

(2) A rectangular block, sides $2a$, $2b$, $2c$,

about an axis through its centre perpendicular to the sides $2a$, $2b$ $= \text{mass} \times \dfrac{a^2 + b^2}{3}$.

(3) An ellipse, semi-axes a and b

about the major axis a $= \text{mass} \times \dfrac{b^2}{4}$,

about the minor axis b $= \text{mass} \times \dfrac{a^2}{4}$,

about an axis through its centre perpendicular to its plane $= \text{mass} \times \dfrac{a^2 + b^2}{4}$.

In the particular case of a circle, radius a, the moment of inertia about a diameter $=$ mass $\times \dfrac{a^2}{4}$, and about a perpendicular to its plane through the centre $=$ mass $\times \dfrac{a^2}{2}$.

(4)　An ellipsoid, semi-axes a, b, c

about the axis a　　　　　　　　　　　　　$=$ mass $\times \dfrac{b^2 + c^2}{5}$.

In the particular case of a sphere of radius a, the moment of inertia about a diameter $=$ mass $\times \dfrac{2a^2}{5}$.

Dr Routh, from whose treatise on Rigid Dynamics this table is taken, gives an easy rule for remembering it.

The moment of inertia about an axis of symmetry

$$= \text{mass} \times \frac{\text{sum of squares of perpendicular semi-axes}}{3, 4, \text{ or } 5}.$$

The denominator is to be 3, 4, or 5 according as the body is rectangular, elliptical (including circular), or ellipsoidal (including spherical).

(5)　A cylinder, radius a, length $2l$,

about its axis　　　　　　　　　　　　　　$=$ mass $\times \dfrac{a^2}{2}$,

about an axis through its centre perpen-
　　dicular to its axis　　　　　　　　　　$=$ mass $\times \left(\dfrac{a^2}{4} + \dfrac{l^2}{3} \right)$.

258.　The usefulness of this table is greatly extended by the following theorem.

The moment of inertia about any axis is equal to the moment of inertia about a parallel axis through the centre of gravity plus the moment of inertia of the whole mass, collected at its centre of gravity, about the original axis.

Let the axis be perpendicular to the plane of the paper at O. Let a parallel axis through the centre of gravity cut this plane in G.

Let $OM = \bar{x}$; $GM = \bar{y}$.

Let m be the mass situated on a line through any point P perpendicular to the plane of the paper.

$$MN = x; \quad PQ = y.$$

Then moment of inertia about axis O

$$= \Sigma m . OP^2 = \Sigma m \left[(\bar{x} + x)^2 + (\bar{y} + y)^2 \right]$$

$$= \Sigma m \left(\bar{x}^2 + \bar{y}^2 + x^2 + y^2 + 2\bar{x}x + 2\bar{y}y \right)$$

$$= \Sigma m . OG^2 + \Sigma m . GP^2 + 2\bar{x}\Sigma mx + 2\bar{y}\Sigma my.$$

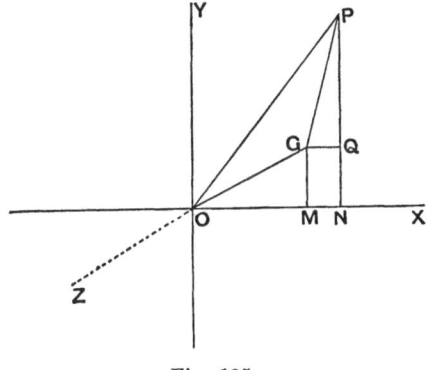

Fig. 135.

But since G is the centre of gravity

$$\Sigma mx = 0 \quad \text{and} \quad \Sigma my = 0, \quad (\S\ 21).$$

And OG is the same for each term of $\Sigma m . OG$.   Therefore

$$\Sigma mOG = OG . \Sigma m = M . OG,$$

if M is the total mass. ΣmGP^2 is the M.I. (Moment of Inertia) about the axis G.

Hence

M.I. about the axis O = M.I. about axis $G + M \times OG^2$.

259. It is easy to see that Moments of Inertia on the one hand and the Statical Moment of the external forces about the axis on the other play the same parts with regard to rotations as masses and impressed forces with respect to motions of translation. Thus for the latter

$$\text{Mass} \times \text{acceleration} = \text{Impressed Force},$$

or $$Ma = P.$$

For the compound pendulum

$$\Sigma mr^2 \times \ddot{\theta} = - Mg . h \sin \theta,$$

i.e., Moment of Inertia × angular acceleration = Moment of Impressed Forces.

In general if I is the moment of inertia, α the angular acceleration, and G the moment of the impressed forces,

$$I\alpha = G.$$

The law expressed by this formula is the exact analogue, for rotations, of the Second Law of Motion, expressed by its formula $Ma = P$, for motions of translation. The one is as fundamental for rotation as the other is for translation.

In the case of constant angular acceleration, kinematical formulae may be found for the angular velocity acquired, ω, and the angle turned through, θ, precisely similar to those for the velocity and distance travelled in § 111.

260. Comparison of formulae for linear and angular motion under constant acceleration.

Kinematical Formulae.

Linear.	Angular.
$v = at$	$\omega = \alpha t$
$s = \dfrac{at^2}{2}$	$\theta = \dfrac{\alpha t^2}{2}$
$\dfrac{v^2}{2} = as$	$\dfrac{\omega^2}{2} = \alpha\theta$

Dynamical Formulae.

$a = \dfrac{P}{M}$	$\alpha = \dfrac{G}{I}$
$Mv = Pt$	$I\omega = Gt$
(Momentum = Impulse)	(Angular Momentum = Moment of Impulse)
$\dfrac{Mv^2}{2} = P \cdot s$	$\dfrac{I\omega^2}{2} = G\theta$
(Energy = Work done)	(Energy = Work done)

The expression $I\omega$, which corresponds to momentum in linear motion, is called *the angular momentum*, or *moment of the momentum* of the rotating body. For the particle m distant r from the axis has a velocity $v = \omega r$ along the tangent. Its momentum is $mv = m\omega r$. The moment of this momentum about the axis is $m\omega r \times r$, and the moment of the momentum for the whole body is $\Sigma m\omega r \times r = \omega \Sigma m r^2 = I \cdot \omega$.

Again, $\dfrac{I\omega^2}{2}$ is the Energy. For the energy of the particle m is

$\dfrac{mv^2}{2} = \frac{1}{2} m\omega^2 r^2$. The energy of the whole body is therefore

$$\Sigma \frac{m\omega^2 r^2}{2} = \frac{\omega^2}{2} \Sigma m r^2 = I \frac{\omega^2}{2}.$$

Lastly, $G\theta$ is the work done by the couple of moment G in turning through an angle whose circular measure is θ.

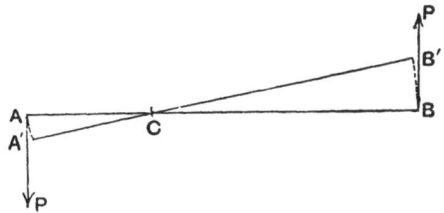

Fig. 136.

For let AB be the arm, and let it turn about any point C through the small angle θ. The work done by the forces is

$$P \times AA' + P \times BB'$$

$$= P \times CA \cdot \theta + P \times CB \cdot \theta$$

$$= P \times AB \cdot \theta$$

$$= \text{moment of couple} \times \theta = G \cdot \theta.$$

261. To make the analogy between the equations of linear and angular motion clearer, let us consider a simple case.

A heavy wheel or disc on an axle O of radius a is set rotating by a cord coiled round the axle and pulled with a steady force P. To find the motion.

Every particle of the wheel must have the same angular acceleration about O. Let this be α. Then by D'Alembert's Principle and taking moments about O

$$- \Sigma \, mar \times r + P . a = 0.$$

$$\therefore \; \alpha = \frac{Pa}{\Sigma mr^2} = \frac{Pa}{I} .$$

Since this is constant, the angular velocity at time t is

$$\omega = \alpha t = \frac{Pa}{I} . t,$$

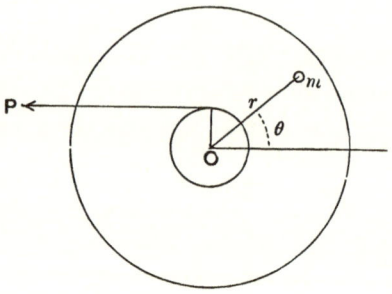

Fig. 137.

and the angle turned through by the wheel from rest is

$$\theta = \frac{1}{2} \frac{Pa}{I} . t^2.$$

At the beginning of Statics we found that, as Leonardo perceived, the *Moment* of a force was the proper measure of its Statical tendency to turn a body round a fulcrum. It now appears that, when rotation ensues, the Moment is still the proper measure of the efficacy of the force in producing angular momentum. So that in all circumstances what we have called the *Torque*, *i.e.*, the twisting or turning effect of a system of forces, is to be measured by their Moment.

262. Next, let the cord, instead of being pulled with constant force, hang vertically and support a mass m. There are now two equations of motion, one for the rotating wheel, and one for the linear motion of m. Let T be the tension of the cord; α the angular acceleration of the wheel; a' the linear acceleration of m.

As before, for the wheel;—

$$\alpha = \frac{Ta}{I} .$$

For m;—

$$a' = \frac{mg - T}{m} .$$

To determine the unknown tension T there is the geometrical relation

$$a\alpha = a'$$

since the acceleration of m is the same as that of the rim of the axle.

Hence
$$\frac{Ta^2}{I} = g - \frac{T}{m},$$

$$\therefore\ T\left(\frac{a^2}{I} + \frac{1}{m}\right) = g, \quad \text{and } T = \frac{mgI}{I + ma^2},$$

so that
$$\alpha = \frac{aT}{I} = \frac{mga}{ma^2 + I},$$

and the motion is known.

EXAMPLES.

1. Calculate in foot-tons the energy of a 10-ton flywheel 8 feet in diameter, revolving 100 times a minute, assuming that the whole mass is collected in the rim.

2. In a certain engine the piston diameter is 8 inches, the mean steam pressure 80 lbs. per square inch, and the length of stroke 3 feet. What must be the mass of a flywheel, supposed collected in the rim, which is to have 6 feet diameter, and store at least 100 times the energy supplied in each stroke when running at 200 revolutions per minute?

3. The inner and outer radii of the rim of a flywheel running at 60 revolutions per minute are 7 and 8 feet respectively, and its mass is 10 tons. Neglecting the spokes, find its energy in foot-tons. How long would it take a 50 H.P. engine to get it up to this speed?

4. In an Atwood's machine the weights are each 8 ounces, the rider 1 ounce, and the pulley weighs 2 ounces. Find the acceleration (1) if the pulley is a ring with spokes of negligible mass; (2) if the pulley is a uniform flat disc.

5. A pendulum consists of a flat rod, 40 inches long and weighing 2 lbs., with a sphere of 6 inches diameter, weighing 10 lbs., rigidly fixed upon it so that the lower end of the rod projects 2 inches beyond the sphere. How many beats will it make per minute?

6. Find the time of swing of a cube about one of its edges which is horizontal and of length $2a$. What must the length of the edge be so that the cube may beat half seconds?

7. The handle of a wheel and axle is let go just as a bucket full of water weighing 60 lbs. reaches the top of a well 18 feet deep, and the bucket gets to the bottom again in 6 seconds. If the axle is 6 inches in diameter, find the moment of inertia of the wheel and axle, neglecting friction.

8. A solid sphere rolls down a rough inclined plane (angle a) without slipping. Shew that its acceleration is $\frac{5}{7} g \sin a$.

9. A hoop rolls down a roof sloping 30° to the horizon for a distance of 16 feet from rest. Shew that its velocity is the same as if it had fallen vertically through 4 feet.

How does this agree with Galileo's principle of motion on an inclined plane ?

10. It has been proposed to draw energy for industrial purposes from the earth's energy of rotation on her axis, by utilising the ebb and flow of the tides. Taking the mass of the earth as $1 \cdot 35 \times 10^{25}$ lbs., and her radius as 4000 miles, shew that, including what is wasted by friction of the tides, she could supply one million horse-power continuously for about $11\frac{1}{2}$ million million years.

CHAPTER XXIX.

EXPERIMENTAL DETERMINATION OF MOMENTS OF INERTIA.

263. THE wheel is mounted on a horizontal or vertical axle
and set in motion by a weight hanging from a
cord coiled round the axle. A loose loop at the
other end of the cord is passed over a pin on the
axle. If the axle is vertical, the cord must be carried over a light
fixed pulley.

Moment of Inertia of a Flywheel.

The work done by the weight in descending a measured
distance to the ground is equal to the energy of the wheel and
of the descending weight, together with what has been expended
in overcoming friction.

Let F be the work used up in overcoming friction during one
turn of the wheel; m the mass of the weight; I the moment of
inertia of the wheel; n the number of revolutions up to the
moment when the weight reaches the ground and the cord slips
off the pin; ω the angular velocity at that moment. Let the
wheel make n' more revolutions before coming to rest.

(1) To find ω. A strip of smoked paper is fastened round
the rim of the wheel by gumming the ends. A tuning fork
making a known number of vibrations per second—say 100—is
held in a clamp, and carries a bristle or metal pointer on one
prong. Just as the weight reaches the ground, the fork is struck
with a rubber cork mounted on a brass wire, and the pointer
pressed lightly for a moment against the smoked paper, by a
slight turn of the clamp on its stand. If the fork is set so as to
vibrate parallel to the axle, the result will be a trace on the

Fig. 138.

smoked paper consisting of a number of waves. Measure the
whole length occupied by as many of these as can be seen

distinctly, and count the waves. Let the length from any point on the first wave to the corresponding point on the pth wave be l. Then the length of one wave is l/p, and since 100 waves pass the fork in one second, the velocity of the rim must have been $100 \times l/p$. Measure the diameter of the wheel by the calipers, and hence find the radius R.

Then $$\omega R = 100 \times l/p \dots\dots\dots\dots\dots\dots(1).$$

Let z be the distance fallen through by the weight. This may be measured before the wheel is allowed to start. Then

$mgz =$ work done by weight

$\quad\quad =$ energy of wheel + energy of weight + loss in friction

$$= I\frac{\omega^2}{2} + m\frac{v^2}{2} + nF\dots\dots\dots\dots\dots\dots\dots\dots\dots\dots\dots\dots(2),$$

where $v =$ the velocity of the weight $= a\omega$.

When the wheel comes to rest after n' more turns, the whole of the work has been absorbed by friction, except the energy of the weight when it was stopped by the ground.

Thus $$mgz = (n + n')F + \frac{ma^2\omega^2}{2} \dots\dots\dots\dots\dots(3).$$

From (3) F may be found, and since a is small, the term $\frac{ma^2\omega^2}{2}$ may generally be neglected, so that

$$F = \frac{mgz}{n + n'},$$

ω is known from (1), and I can be found from (2).

As a check on the counting, see whether $z = 2\pi a . n$. It is better to measure z directly, instead of calculating it from this formula, to avoid errors arising from uneven winding and stretching of the string.

264. A heavy cylinder is clamped to one end of a brass
The Torsion   wire. The other end of the wire is held in
Pendulum.   a fixed clamp.

The cylinder carries a cross arm, to the ends of which small weights may be attached, and a pointer moving over a flat circular scale, so that the angle turned through by the cylinder can be read.

When the cylinder is turned about the vertical wire, the twisted wire exerts a couple tending to bring it back to the original position. The moment of this couple is proportional to the angle turned through. (§ 210.)

Turn the cylinder through a small angle and set it free.

Let θ be the angular deflection at any moment; $\ddot{\theta}$ the angular acceleration; I the M.I. of the suspended system about the vertical.

Then by D'Alembert's Principle, as in the case of the Compound Pendulum,

Fig. 139.

M.I. × angular acceleration $= -$ (moment of twisting couple).

Let G be the torque, or moment of the couple required to twist the wire through a unit angle. Then $G\theta$ is the couple for a deflection θ.

$$\therefore \; I\ddot{\theta} = -G \cdot \theta,$$

$$\ddot{\theta} = -\frac{G}{I} \cdot \theta,$$

and the motion will be Simple Harmonic with a period $2\pi\sqrt{\dfrac{I}{G}}$.

In fact the best proof that the couple is proportional to the angle of torsion is that the vibrations are simple harmonic vibrations. Verify that this is so, by timing 20 swings for amplitudes

of 5°, 10°, 20°, 30° on each side of zero.   The time of swing will
be found to be the same.

Observe accurately the time of 50 complete oscillations.   It is
best to note the instants of passing the zero, not the moment of
coming to rest on one side or the other, since the exact moment
can be more sharply fixed when the pointer is moving rapidly.

From the duration of 50 swings find the time of one swing.
Let this be T_1.   Then

$$T_1 = 2\pi \sqrt{\frac{I}{G}} \quad\ldots\ldots\ldots\ldots\ldots\ldots\ldots(1).$$

Hang from the ends of the cross arm two equal small weights,
having first found their masses, m, by weighing them on a sensitive
balance.

Let T_2 be the time of swing with the weights added.   Measure
l the distance of either small weight from the vertical axis.   The
moment of inertia of the system is now $I + 2ml^2$; the couple G,
depending only on the wire, is unaltered.   So that

$$T_2 = 2\pi \sqrt{\frac{I + 2ml^2}{G}} \quad\ldots\ldots\ldots\ldots\ldots(2).$$

Equations (1) and (2) determine I and G.

Thus
$$\frac{I + 2ml^2}{I} = \frac{T_2^2}{T_1^2},$$

$$\frac{2ml^2}{I} = \frac{T_2^2 - T_1^2}{T_1^2},$$

$$I = 2ml^2 \cdot \frac{T_1^2}{T_2^2 - T_1^2}.$$

The difference $T_2^2 - T_1^2$ must not be very small, or a slight
error in determining T_1 and T_2 will make a great difference in the
value of I.   If, however, larger weights are employed, they cannot
be treated as particles, and their moment of inertia about the axis
must be expressed by adding to $2ml^2$ the sum of their moments
of inertia about vertical axes through their own centres of gravity.

265.   From (1)

$$G = \frac{4\pi^2}{T_1^2} I.$$

Having once found G for a particular wire, we may use the wire to determine by a single observation the moments of inertia of any objects that can be clamped to it. For if K be the moment of inertia of the object, the time of vibration when it is suspended by the wire will be $T = 2\pi \sqrt{\dfrac{K}{G}}$. Observing T we can at once find K.

The student should in this manner verify some of the formulae in § 257.

266. The important constant called the modulus of torsion is the moment of the couple required to twist a unit length of the wire through a unit angle, *i.e.* one radian. Let this be denoted by τ. Then the couple required to twist one end of a wire of length l through an angle θ will be $\dfrac{\tau}{l}\,\theta$, since each unit of length is only twisted through $\dfrac{\theta}{l}$.

Modulus of Torsion of a wire by Maxwell's Needle.

We might measure the length of the wire in experiment § 264, and determine τ from the equation

$$\frac{\tau}{l} = G.$$

But it may be found much more accurately by means of a piece of apparatus devised by Clerk Maxwell for use in the study of the viscosity of gases. Figure 140 shews the instrument, which is known as Maxwell's Needle.

The "needle" is a hollow brass cylinder provided with a central clamp for suspension by the wire. The other end of the wire is held in a clamp supported at the head of a hollow vertical brass pillar, with a torsion head, *i.e.* the top of the pillar, bearing the clamp, can be turned round so as to adjust the position of equilibrium of the needle. There is a glass case, supported on levelling screws, to protect the needle from disturbance by currents of air.

Inside the needle slide four brass cylinders, each one-quarter of its length. Two of these are hollow, and of equal weight; the other two are filled with lead, and are also of equal weight. There

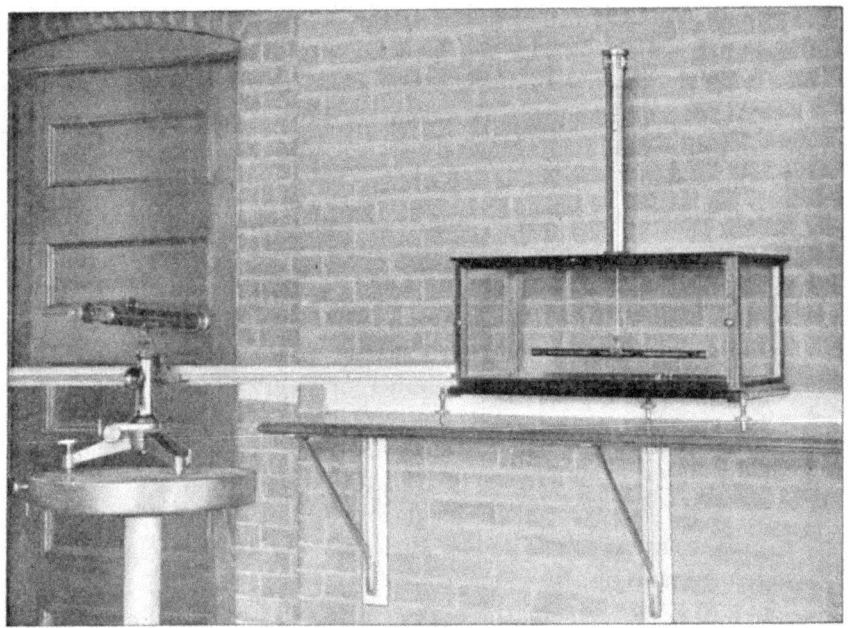

Fig. 140.

is a scale on the needle by which the inner cylinders can be set accurately in position.

To find the modulus of torsion of a wire, the needle is suspended by it, and the cylinders are slid into position with the two heavy ones on the inside, and the hollow ones at the ends. The needle is set vibrating, and the time of oscillation T_1 is determined.

The cylinders in the needle are then rearranged so that the heavy ones are at the ends; and the time of oscillation T_2 is again determined.

To assist in finding T_1 and T_2 with the greatest accuracy the needle is provided with a small plane mirror at its centre. Opposite this a reading telescope with scale is set up so that the image of the scale formed by the mirror is seen sharply in the telescope. The torsion head can be turned till the zero at the middle of the image of the scale coincides with a vertical cross wire in the focus of the telescope. When the needle is oscillating, the scale will

appear to cross the field of view of the telescope, and the instant
of the passage of the zero can be fixed with great exactness. By
observing the time of 10 swings, taking the time from a laboratory
clock ticking seconds, or a chronometer ticking half seconds, the
time of swing may be first found to within ·05 of a second. From
this the approximate time at which the hundredth swing will
occur is calculated. The needle is left swinging, and when the
proper moment approaches, the observer takes his station and
records the actual instant of the hundredth passage. From this
a much closer value of the time of swing may be found; and if
necessary, this may be employed in the same way to allow the
observation of a still greater number of swings.

267. Let m, m' be the masses of the loaded and hollow
cylinders respectively; c the length of a cylinder;
I the M.I. of the empty needle; I_1 and I_2 those of
the loaded and hollow cylinders about vertical axes through their
centres of gravity.

Theory of the Instrument.

Let K be the M.I. of the system with the loaded cylinders in
the middle; $K + k$ the M.I. when the heavy cylinders are at the
ends.

Then
$$T_1 = 2\pi \sqrt{K \Big/ \frac{\tau}{l}},$$

$$T_2 = 2\pi \sqrt{(K + k) \Big/ \frac{\tau}{l}}.$$

Thus
$$K \cdot \frac{l}{\tau} = \frac{T_1^2}{4\pi^2}; \quad (K + k) \cdot \frac{l}{\tau} = \frac{T_2^2}{4\pi^2}.$$

$$\therefore k \cdot \frac{l}{\tau} = \frac{T_2^2 - T_1^2}{4\pi^2},$$

and
$$\tau = \frac{4\pi^2 kl}{T_2^2 - T_1^2}.$$

It remains to find k. By § 258
$$K = I + 2I_1 + 2I_2 + 2m \left(\frac{c}{2}\right)^2 + 2m' \left(\frac{3c}{2}\right)^2,$$

and
$$K + k = I + 2I_1 + 2I_2 + 2m \left(\frac{3c}{2}\right)^2 + 2m' \left(\frac{c}{2}\right)^2.$$

$$\therefore k = \frac{c^2}{4} \cdot (m - m'),$$

whence
$$\tau = \frac{\pi^2 c^2 (m - m') l}{T_2^2 - T_1^2}.$$

The length l of the wire from clamp to clamp is measured with a beam compass or steel tape.

Having found this constant for a particular wire, Maxwell employed the wire for suspending flat discs in gases at various pressures, and from the observed damping of their oscillations he deduced the effects of the viscosity of the gases on the surfaces of the discs.

268. The method of altering the M.I. by adding or shifting bodies of known mass and form, whose moments of inertia can be calculated, and then observing the time of swing, is employed in many physical measurements. Thus in determining the strength of the Earth's magnetic field by the Kew pattern of magnetometer, the M.I. of the suspended magnet is found by adding to it a small brass cylinder. As brass is non-magnetic, the couple exerted on the system by the earth's field is not altered.

The M.I. of the magnet in a Ballistic Galvanometer is sometimes found in the same manner, but more usually by the method employed with the Ballistic Pendulum of the following article.

269. The Ballistic Pendulum figured on p. 298 is merely a massively constructed Balance. The heavy framework representing the beam rests on steel knife edges in steel V-shaped cups. The heavy weight hanging below is the gravity bob for adjusting the distance of the centre of gravity from the knife edges. Above are seen a large weight and a small metal flag for coarse and fine adjustment of the pointer to zero. In this instrument the readings may also be taken by means of the image of the curved scale seen against a horizontal fixed pointer.

Determination of the velocity of a bullet by the Ballistic Pendulum.

The two heavy cylinders suspended from the beam may be filled with shot and adjusted to equal weights. They hang by knife edges resting in grooves which serve to graduate the beam. They can thus be set at equal measured distances from the axis, and as they do not rotate with the beam, but always hang

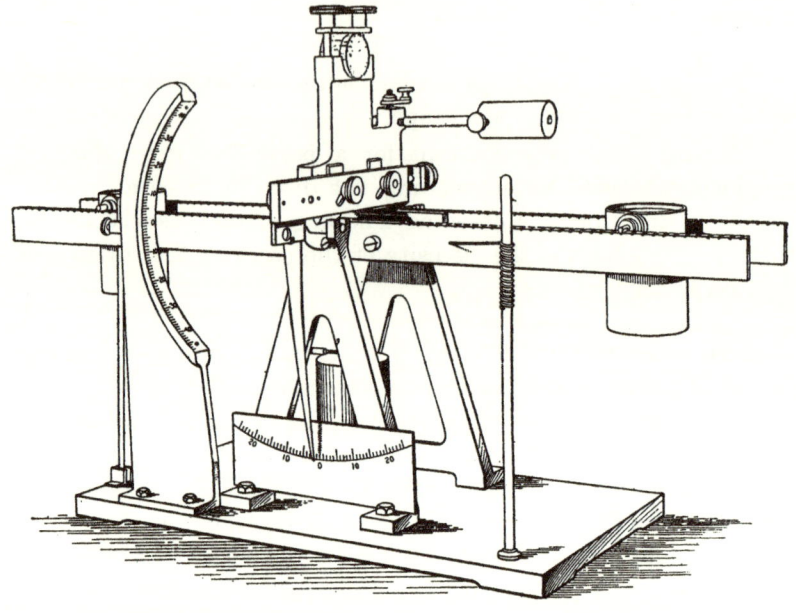

Fig. 141.

vertically, the change they make in its M.I. is merely the product of their masses into the squares of the distances of their points of suspension from the central knife edge.

Moreover, a horizontal shift of these weights does not change the distance of the C.G. from the axis, so that the couple tending to restore equilibrium is unaltered. They thus afford a means of both measuring the M.I. and adjusting it to a convenient value, without otherwise changing the circumstances.

Above the centre of the beam is an upright carrying a heavy metal disc on which a blow may be struck; or a block of wood may be clamped there to receive a bullet from a revolver. Such a blow will cause the instrument, previously at rest, to swing through a certain angle, and then continue oscillating. The object is to find from its movements the momentum communicated to it by the blow. From this the velocity of the striking object may be found if its mass is known.

270. Let I be the M.I. of the pendulum about the knife edges.

Theory of the Instrument. ω = the angular velocity with which it begins to swing.

m = the mass of the bullet, and v its velocity.

k = distance from knife edge to line of fire of bullet.

Then

Moment of momentum of pendulum

= moment of impulse of bullet about the axis.

$$I\omega = mvk,$$

and $$v = \frac{I\omega}{mk} \quad \dots\dots\dots\dots\dots\dots\dots\dots\dots\dots\dots(1).$$

(Strictly, I, the moment of momentum of the pendulum, should include that of the bullet embedded in it. This may be partially allowed for by taking the time of swing after the bullet has been fired. But in any case the mass of the bullet is too small to make any appreciable difference.)

(1) To find I.

Let M be the mass of the pendulum; $OG = h$ the distance of the C.G. from the axis.

Observe the time of oscillation of the pendulum. Let it be T.

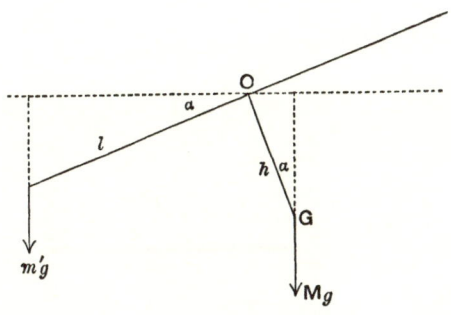

Fig. 142.

Then

$$T = 2\pi \sqrt{\frac{I}{Mgh}} \quad \dots\dots\dots\dots\dots\dots(2).$$

Next, place a small weight of mass m' in one of the cylinders; let l be the distance of a cylinder from the axis, and let α be the steady deflection from the horizontal when the pendulum comes to rest.

Then by moments about O

$$Mg \cdot h \sin \alpha = m'g \cdot l \cos \alpha \quad \dots\dots\dots\dots\dots(3).$$

From (2) and (3),

$$I = \frac{T^2}{4\pi^2} \cdot Mgh = \frac{T^2}{4\pi^2} \cdot m'gl \cot \alpha.$$

(2) To find ω.

The pendulum will swing aside when the blow is struck till its energy is exhausted by the work done against gravity. Let β be the angle through which it swings before it comes to rest for the *first* time.

The c.g. rises through a vertical height

$$GH = OG - OH = h\,(1 - \cos \beta).$$

The work done against gravity
$$= Mg \cdot h\,(1 - \cos \beta),$$

$$\therefore \frac{I\omega^2}{2} = Mgh\,(1 - \cos \beta)$$

$$= 2Mgh \sin^2 \frac{\beta}{2},$$

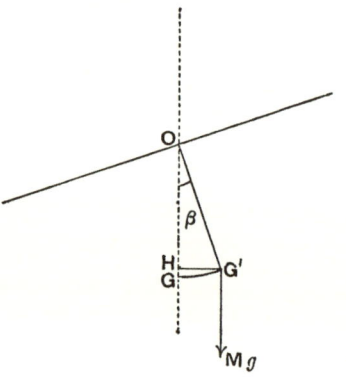

Fig. 143.

and $\qquad \omega = 2\sqrt{\dfrac{Mgh}{I}} \cdot \sin \dfrac{\beta}{2} = \dfrac{4\pi}{T} \sin \dfrac{\beta}{2}$, by (2).

Hence finally

$$v = \frac{I\omega}{mk} = \frac{T^2}{4\pi^2} \frac{m'gl \cot \alpha \times \dfrac{4\pi}{T} \cdot \sin \dfrac{\beta}{2}}{mk}$$

$$= \frac{m'gl \cdot T}{m\pi k} \cdot \cot \alpha \cdot \sin \frac{\beta}{2}.$$

271. The experiment should be conducted in the following order.

Adjust the balance to zero after the wooden block is clamped on.

Observe :—

(1) The steady deflection α when the known weight m' is placed in one of the cylinders, using both the pointer and the reflected scale.

(2) The distance of cylinder-suspension from the knife edge $= l$.

(3) The value of the first swing when the bullet strikes $= \beta$. Be sure that the Balance is at rest before firing. Practise observing the first swing when a blow is struck, before trying with the bullet.

(4) The distance from axis of line of fire of bullet $= k$.

(5) The time of vibration after the bullet has been fired in $= T$.

(6) The mass of the bullet $= m$.

From these observed values v is determined by the formula.

272. The moment of inertia I might have been found by shifting the two cylinders to another distance l' from the axis, and again observing the time of vibration T'. Weigh one of the cylinders, and let its mass be M'. Then, if K be the M.I. of the frame without the cylinders,

$$T = 2\pi \sqrt{\frac{K + 2M'l^2}{Mgh}},$$

$$T' = 2\pi \sqrt{\frac{K + 2M'l'^2}{Mgh}},$$

whence K is known, and $I = K + 2M'l^2$, may be calculated.

We have preferred to use the steady deflection by a known weight, because this method is a strict mechanical analogue to the practice with the Ballistic Galvanometer in electrical measurements.

EXAMPLES.

1. The following observations are made with the magnet of a Kew Magnetometer :—

Time of swing without brass cylinder, 5·055 sec.

Time with brass cylinder, axis of cylinder perpendicular to axis of suspension, 8·746 sec.

Dimensions of cylinder,

$$\text{length} = 9·548 \text{ cms.}$$
$$\text{diameter} = 0·998 \text{ cms.}$$
$$\text{weight} = 63·38 \text{ gms.}$$

Hence find the moment of inertia of the magnet.

2. Determine the velocity of a revolver bullet from the following observations with the Ballistic Pendulum :—

50 gms. placed in one cylinder caused a deflection of 14° 20′.

Distance between cylinder-suspensions = $2l$ = 80 cms.

First swing = 13° 40′.

Distance of line of fire from knife edge = 14·2 cms.

Time of vibration = 6·7 seconds.

Mass of bullet = 88 grains.

The revolver used in this experiment was a 32-calibre Ivor and Johnson with a 10 grain charge of powder.

CHAPTER XXX.

DETERMINATION OF THE VALUE OF GRAVITY BY KATER'S PENDULUM.

273. THE acceleration produced by gravity at any place is a physical constant of great importance. By far the best means of finding it is the pendulum (§ 223). But if a simple pendulum is used, no great accuracy is attainable. The time of swing cannot be exactly found unless the pendulum can make many hundred swings before the arc, which must be small to begin with, becomes too short for observation. For this purpose the bob must be heavy, and this requires a strong wire or string to support it. The pendulum cannot then be treated as an ideal simple pendulum, and yet it is not possible to allow for the mass of the string, or to fix the position of the centre of gravity of the bob. The distance from the point of suspension to the centre of the bob cannot be measured with the utmost refinement while it is in position, yet if it is taken down, the string is no longer stretched to the same length.

Captain Kater, a member of the Committee appointed by the Royal Society to determine as accurately as possible the length of the Seconds Pendulum at Greenwich, after many failures with the simple pendulum, was casting about for some property of the pendulum which would enable him to overcome these difficulties, when he hit upon a discovery by Huyghens concerning the Compound Pendulum which answered all requirements. The paper describing his experiments is in the *Philosophical Transactions* for 1818.

274. Let C be the centre of suspension of a pendulum of any shape; G its centre of gravity; CO the length of the simple equivalent pendulum. The point O was called by Huyghens the *Centre of Oscillation* for the axis C; and he shewed that if the body were suspended from a parallel axis through O, the time of oscillation would be the same, and the length of the simple equivalent pendulum would be the same; so that C would become the centre of oscillation for the axis O. C and O, the centres of suspension and oscillation, are thus convertible, and *the time of swing about each is the same as that of an ideal simple pendulum of length CO.*

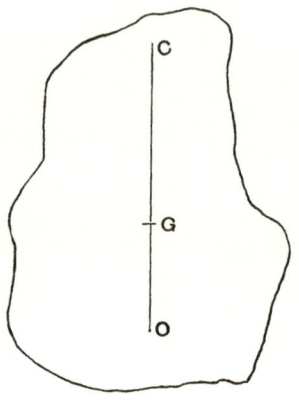

Fig. 144.

The proof is as follows:

Let M be the mass of the pendulum, I its M.I. about the centre of gravity G; and let $CG = h$; $CO = l$; $OG = h'$.

Then

M. I. of the pendulum about axis $C = I + Mh^2$,

and M. I. „ „ „ $O = I + Mh'^2$.

The time of swing about C is that of the simple equivalent pendulum,

$$T_1 = 2\pi \sqrt{\frac{I + Mh^2}{Mgh}} = 2\pi \sqrt{\frac{l}{g}};$$

$$\therefore \frac{I + Mh^2}{Mh} = l = h + h'; \quad \therefore I = Mhh'.$$

The time of swing about O is

$$T_2 = 2\pi \sqrt{\frac{I + Mh'^2}{Mgh'}}$$

$$= 2\pi \sqrt{\frac{Mhh' + Mh'^2}{Mgh'}} = 2\pi \sqrt{\frac{h + h'}{g}}$$

$$= 2\pi \sqrt{\frac{l}{g}} = T_1.$$

275. Drill a hole through a board of any shape. Pass a knitting needle through the hole, and adjust a simple pendulum, consisting of a bullet on a string, as in § 250, till the time of swing is the same for both. Mark the point behind the bullet, and pass the needle through a hole drilled at the mark, without altering the length of the string. It will be found that the board still swings with the bullet.

Two other curious properties of the point O may be mentioned here. Let us suppose that instead of the board, a cricket bat is suspended on an axis C passed through the handle where it is grasped by the hands. The corresponding centre of oscillation O is the point of the bat with which a ball should be struck so as to produce the greatest effect on the ball with a given swing. It is also the point with which the ball must be struck so that there shall be no unpleasant jar at the hands. Every cricketer knows that when he makes his best hits he does not feel the blow of the ball at all. For these reasons the Centre of Oscillation is also called the *Centre of Percussion*.

The reason for these properties may be understood without equations. Imagine the simple equivalent pendulum to consist of a heavy mass suspended, not by a string, but by a rigid, weightless rod, something like the 16 lb. hammer on its handle. Then for motion about the hands the bat and the pendulum are dynamically similar, and we may infer the properties of the one from those of the other.

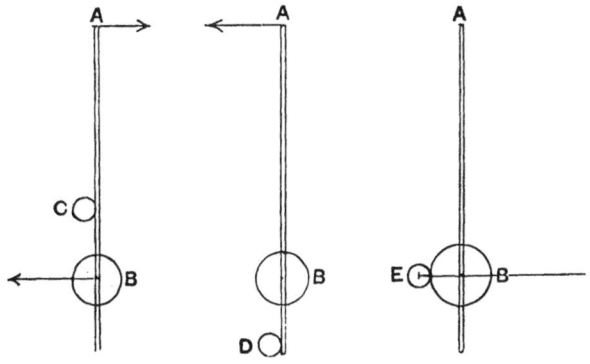

Fig. 145.

If the swinging hammer strike an obstacle at C, on the handle above the head B, the momentum of the head carries it on round C as a fulcrum, jerking the hands at A backwards, and being partly wasted in producing the jerk.

If the obstacle meets the handle at D, below the head, the momentum of B jerks A forwards, and is again partly wasted.

Only when B strikes the obstacle directly, as at E, is the whole momentum given up to it, as if the balls were free of the handle. And then there is no jerk at A.

The same considerations hold for the bat, which moves as if its whole mass were concentrated at the centre of oscillation or percussion O. Verify this by suspending the board-pendulum by a light thread instead of the knitting needle. Hold the thread horizontally with the pendulum at rest and let an assistant strike a blow on its edge along a horizontal line through the centre of percussion, previously determined by aid of the bullet and string. It will be found that quite a smart blow may be struck without breaking the thread. The board begins to turn of its own accord about the thread. But if the line of the blow be not exactly through the centre of percussion, a slight tap will suffice to break the thread. Instead of the board it is better to use a straight flat rod suspended by a hole near one end. Its centre of percussion is at two-thirds of its length from the top.

276. The property of the Centre of Oscillation discovered by Huyghens affords the means of constructing the equivalent of an Ideal Simple Pendulum, and of measuring with great exactness both its length and the time of swing. To apply it, Kater constructed a pendulum of a bar of plate brass, $1\frac{1}{2}$ in. wide and $\frac{1}{8}$ in. thick. Two knife edges of hard steel were ground true and firmly fixed near the ends at a distance of about 39·4 inches apart. A brass weight of 2 lbs. 7 oz. was fixed between one end of the bar and the nearer knife edge; a second weight of $7\frac{1}{2}$ oz. was made to slide on the bar near the other knife edge and between them; and a small slider of 4 oz., capable of fine adjustment by a screw, was placed near the middle of the bar.

The pendulum could be suspended by either knife edge resting upon polished agate planes let into a firm support. The times of vibration about the two knife edges were roughly found; the second weight of 7½ oz. was moved along the bar till they were nearly equal; and then the slider was carefully adjusted till the time of vibration about either knife edge was practically the same. This would therefore be the time of vibration of an ideal pendulum of a length equal to the distance between the knife edges. It remained to find this distance and the time of swing accurately.

277. The pendulum was mounted on a comparator
Measurement of   under reading microscopes which were
the Length.   brought over the knife edges, and compared with standard scales. The mean results of three sets of measurements taken at different times were within one ten-thousandth of an inch of each other. The corrected mean of all the sets was

$$39\cdot44085 \text{ inches at } 62° \text{ F.}$$

278. The pendulum was compared with an astro-
Determination of   nomical clock beating seconds, and the
the Time of   rate of the clock was found from transit   Fig. 146.
Swing.   observations taken every day during the
experiments, and often more frequently.

It is not possible to observe the exact time of a chosen number of vibrations, for this would require the fraction of a second to be estimated either by the eye or ear. Nor could the number of vibrations in a fixed interval be observed, for this would require a fraction of a vibration to be estimated. The device by which this difficulty was avoided is known as the *Method of Coincidences.* The pendulum was mounted on a firm support in front of the clock. (The second clock in the figure was only used as a check.)

A small disc of white paper was fixed on the bob of the clock pendulum, of the same width as the two slips of blackened deal projecting from the ends of the Kater pendulum. At a distance of nine feet in front of the clock was a telescope with the field of

view limited by a diaphragm to a narrow vertical slit exactly wide
enough to shew one of the deal slips or the paper disc. If both
pendulums were at rest, the disc would be just hidden by the
deal slip.

The distance between the knife edges had been fixed at about
39·4 in. so that the period of vibration should be very nearly that

Fig. 147.

of the clock pendulum, but a little greater. "If both pendulums be now set in motion, the brass pendulum a little preceding that of the clock, the slip of deal will first pass through the field of view of the telescope at each vibration and will be followed by the white disc. But the clock will gain on the pendulum, so that the white disc will gradually approach the slip of deal, and at a certain vibration will be wholly concealed by it. The minute and second at which this total disappearance is observed must be noted. The pendulums will now be seen to separate, and after a time will again approach each other, when the same phenomenon will take place. The interval between the two *coincidences* in seconds will give the number of vibrations made by the clock; and the brass pendulum must have made two less than the clock. Hence by simple proportion the time of vibration of the pendulum is known." Kater calculated the number of vibrations made in 24 hours, or 86,400 seconds.

To shew the accuracy attainable by this method, let us suppose an error of one whole second in observing a coincidence (it will be noticed that no fractions have to be estimated). The interval between coincidences in Kater's experiments was about 530 seconds. With this value the pendulum must have made 528 swings, so that its time of swing would be 530/528 seconds, *i.e.* 1·003787 seconds. If by error the interval were taken as 531 seconds, the time would be $531/529 = 1\cdot0037807$ seconds. This does not differ from the true value by one part in 100,000.

279. The observed result must be corrected for

(1) Expansion or contraction of the pendulum; due to difference of temperature at the time of the experiment from the temperature, 62° F., at which the length between knife edges was measured.

(2) Variation from Simple Harmonic Motion due to the size of the arc. The arc was read by a whalebone pointer moving over a scale of degrees. This correction is called the "reduction to infinitely small arcs."

(3) Buoyancy of the air, making the weight different from what it would be in vacuo. This depended on the height of the barometer.

280. Twelve sets of observations similar to Table I. were made. The results are summarized in Tables II. and III. In Table II. it will be observed how nearly the time of swing was the same for the two positions of the pendulum.

TABLE I.

	Temp.	Time of co-incidence	Arc of vibration	Mean arc	Interv. in seconds	No. of vibrats.	Vibrations in 24 hours	Corr. for arc	Vibrations in 24 hours
Slider 19 divisions Clock gaining 0″,18 on mean time			GREAT WEIGHT *above*						Barometer 29,90
July 3rd	° 68,3	m. s. 45. 3 53.27 1.51 10.16	° 1,23 1,03 0,87 0,74	° 1,13 0,95 0,80	504 504 505	502 502 503	86057,16 86057,16 86057,82	s. 2,08 1,47 1,04	86059,24 86058,63 86058,86
	68,4	18.42	0,63	0,68	506	504	86058,49	0,75	86059,24
								Mean Clock	86058,99 + 0,18
	68,4	mean							86059,17
M			GREAT WEIGHT *below*						
	68,4	24.31 32.54 41.18 49.42	1,24 1,11 0,99 0,90	1,17 1,05 0,94	503 504 504	501 502 502	86056,47 86057,16 86057,16	2,23 1,80 1,44	86058,70 86058,96 86058,60
	68,5	58. 8	0,82	0,86	506	504	86058,49	1,20	86059,69
								Mean Clock Temp.	86058,99 + 0,18 + 0,04
	68,5	mean							86059,21

The results of such of the preceding experiments as are to be used for calculating the length of the seconds pendulum, are brought under one view in the following table:

TABLE II.

Place of the slider	Expt.	Temp.	Barom.	No. of vibrations. Great wt. *above*	Diff.	No. of vibrations. Great wt. *below*	Vibs. in excess or defect
23	A	68,7	29,76	86059,39	,03	86059,42	–
23	B	71,3	29,86	86057,70	,23	86057,93	–
23	C	71,4	29,86	86057,93	,23	86057,70	+
23	D	73,1	29,95	86056,54	,43	86056,97	–
Pendulum re-measured							
21	E	69,3	29,70	86058,88	,06	86058,94	–
20	F	69,3	29,70	86058,89	,12	86059,01	–
20	G	68,5	29,70	86059,03	,19	86059,22	–
18	H	68,7	29,70	86059,36	,11	86059,25	+
18	I	69,3	29,70	86059,19	,16	86058,93	+
18	K	69,3	29,70	86059,14	,31	86058,83	+
19	L	68,1	29,90	86059,26	,04	86059,22	+
19	M	68,4	29,90	86059,17	,04	86059,21	–
			Mean	86058,71		86058,72	

" Using the vibrations when the great weight was *below*, as being nearer to the truth than in the other position of the pendulum, we obtain the following results."

TABLE III.

Expt.	Temp.	Barom.	Vibrations in 24 hours	Length of the seconds pen. in air	Corr. for the atmosphere	Length of the seconds pend. in vacuo	Difference from the mean
A	68,7	29,76	86059,42	39,13313	,00544	39,13857	+ ,00028
B	71,3	29,86	86059,93	39,13278	,00544	39,13822	– ,00007
C	71,4	29,86	86057,70	39,13260	,00544	39,13804	– ,00025
D	73,1	29,95	86056,97	39,13259	,00544	39,13803	– ,00026
E	69,3	29,70	86058,94	39,13293	,00544	39,13837	+ ,00008
F	69,3	29,70	86059,01	39,13298	,00544	39,13842	+ ,00013
G	68,5	29,70	86059,22	39,13286	,00545	39,13831	+ ,00002
H	68,7	29,70	86059,25	39,13296	,00544	39,13840	+ ,00011
I	69,3	29,70	86058,93	39,13291	,00544	39,13834	+ ,00005
K	69,3	29,70	86058,83	39,13282	,00544	39,13825	– ,00003
L	68,1	29,90	86059,22	39,13271	,00548	39,13819	– ,00009
M	68,4	29,90	86059,21	39,13281	,00548	39,13829	– ,00000
					Mean	39,13829	

The mean value was then corrected for the height of Portland Place, where the observations were made, above the sea level. The final result was

39·1386 inches

for the length of the pendulum vibrating seconds at the level of the sea in the latitude of London.

EXAMPLE.

From Kater's value of the length of the seconds pendulum determine the value of gravity at the sea-level in London.

CHAPTER XXXI.

THE CONSTANT OF GRAVITATION, OR WEIGHING THE EARTH. THE CAVENDISH EXPERIMENT.

281. JUST twenty years before Kater determined accurately the value of the acceleration of gravity at the earth's surface, Henry Cavendish succeeded in measuring another important physical constant, which was necessary to complete the statement of the law of universal gravitation.

According to Newton's Law, the force of attraction between any two masses is directly proportional to the product of the masses and inversely proportional to the square of their distance. This enables us to *compare* attractions, but before we can calculate the actual value of the attraction between two given masses in units of force otherwise known to us, we must, as in every case of proportion, know what it is in some standard instance.

Suppose, for example, we agree to measure masses, distances, and forces in grammes, centimetres, and dynes, and *know* that the attraction exerted by a particle of mass one gramme on an equal particle distant one centimetre from it is G dynes. Then:

Attraction of 1 gm. on 1 gm. at 1 cm. is					G		dynes
,,	,, m_1	,,	1	,,	1	,,	Gm_1	,,
,,	,, m_1	,,	m_2	,,	1	,,	Gm_1m_2	,,
,,	,, m_1	,,	m_2	,,	d	,,	$\dfrac{Gm_1m_2}{d^2}$	,,

G is called the constant of Gravitation. Its value, like that of the dyne and the poundal, can only be found by an experiment. If we chose pounds, feet, and poundals for our units, G would of course have a different numerical value.

282. From another point of view the determination of G is equivalent to finding the mass of the earth in terms of the ordinary units of mass.

Newton had been able to compare the mass of the Earth, or of any planet *which had a satellite*, with that of the Sun. For let S, E, M be the masses of the Sun, Earth, and Moon; let R, r be the radii of the Earth's and Moon's orbits, supposed circular; and let T_e, T_m be the times of revolution of the Earth and Moon in their orbits. Then (§ 77),

acceleration of Earth to Sun $= \dfrac{4\pi^2}{T_e{}^2} . R,$

and acceleration of Moon to Earth $= \dfrac{4\pi^2}{T_m{}^2} . r.$

But, by the law of gravitation, force exerted by Sun on Earth $= G . \dfrac{S . E}{R^2} .$

Therefore, by Second Law of Motion, acceleration of Earth to Sun $= \dfrac{G . S . E}{R^2} = G . \dfrac{S}{R^2}.$

Similarly, acceleration of Moon to Earth $= G . \dfrac{E}{r^2}.$

Therefore $\dfrac{4\pi^2}{T_e{}^2} . R = G . \dfrac{S}{R^2},$

and $\dfrac{4\pi^2}{T_m{}^2} . r = G . \dfrac{E}{r^2};$

whence $\dfrac{E}{S} = \dfrac{r^3}{R^3} . \dfrac{T_e{}^2}{T_m{}^2},$

in which R, r, T_e, T_m are known from observation.

But this does not tell us what the mass of the Earth or the Sun is in pounds or grammes.

If, however, we knew the value of G in dynes, then, since the weight of a gramme is the attraction exerted on it by the whole mass of the Earth, supposed collected at its centre (§ 238), we should have

$$G . \dfrac{E . 1}{a^2} = \text{weight of 1 gm. at the surface of the Earth}$$

$$= 981 \text{ dynes};$$

whence
$$E = \frac{981}{G} \cdot a^2,$$

and E is given in grammes when a, the radius, is expressed in centimetres.

The Earth would thus be "weighed," *i.e.* expressed in terms of a known unit of mass.   Or, we could express Δ, the mean density of the Earth, compared with water, since

$$E = \tfrac{4}{3}\pi a^3 \cdot \Delta \text{ gms.},$$

so that Δ is known, if E is known.

283.   The problem is thus reduced to finding by an experiment the actual attraction in dynes between two known masses.

Now the attraction of the whole earth on one gramme only amounts to the weight of one gramme.   Thus the attraction between any two masses accessible to us must be a very minute force.

Two lines of attack are open to us.   Either we must choose for one of the masses some large natural mass, such as a mountain, in the hope of increasing the force up to a measurable magnitude; or else apparatus of extreme sensitiveness must be devised, if we are to detect the attraction between such masses as can be dealt with in a laboratory.

284.   The first method was tried by Bouguer in 1740, on Chimborazo, a mountain 20,000 feet high in the Andes; and again, with great care, by Maskelyne in 1772, on Schehallion in Scotland. Each of these observers suspended a plumb line on the side of the mountain, and noted the deflection from the true vertical, which could be determined astronomically.   Similar experiments have since been carried out at Arthur's Seat near Edinburgh.

In 1854 Airy, another English Astronomer Royal, observed the change in the time of swing of a pendulum when it was removed from the surface of the earth to the bottom of the Harton coal pit, 1250 feet deep.   Part of this was to be attributed to change of distance from the centre, and could be calculated.   The rest was due to the removal of a layer of the earth's crust 1250 feet thick, since the outer shell has no attraction on the pendulum (§ 237). Others (Carlini in 1821 on Mont Cenis, and Mendenhall in 1880 on Fujïyama in Japan) made similar determinations by removing the pendulum to the top of a high mountain.

The objection to all these methods is that the exact shape, and above all the density of the rocks and strata in the disturbing mass cannot be accurately determined.

285. The first to plan an experiment on the laboratory method was the Rev. John Michell, who constructed the apparatus, but did not live to make the experiment. His apparatus passed to Professor Wollaston of Cambridge, who handed it over to Henry Cavendish, noted for his skill as an experimenter in electricity and other branches of Physics.

In carrying out the famous experiment which goes by his name, Cavendish retained Michell's original idea, and even the dimensions of the apparatus; but found it advisable to reconstruct most of it. The account of his work was communicated to the Royal Society in 1798, and is to be found in the volume of the *Philosophical Transactions* for that year.

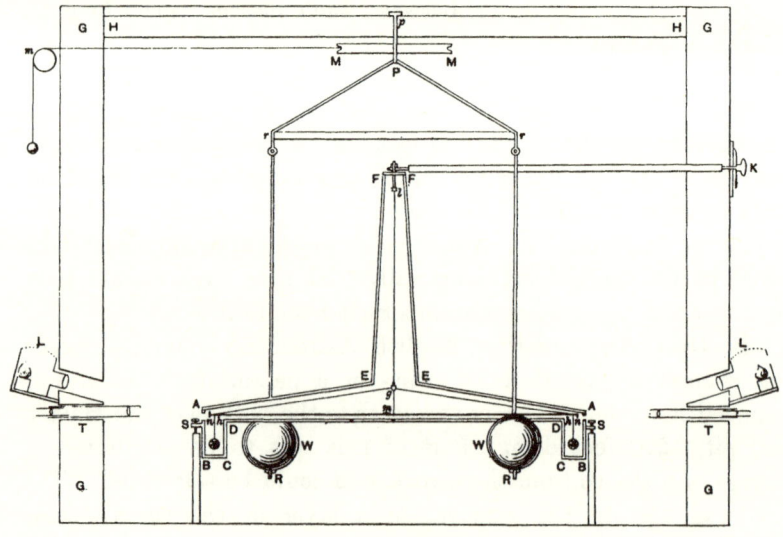

Fig. 148.

286. The apparatus was essentially a Torsion Pendulum, similar to that of § 264. About the same time Coulomb independently applied this instrument to the measurement of small electric and magnetic attractions.

It consisted of a slender deal rod *hh*, 6 feet long, stiffened by an upright *gm* braced with a silver wire *hgh*, and suspended by a fine silvered copper wire 39¼ inches long. The wire was fastened by clamps to the upright on the rod and to the torsion head *FF*. The balls to be attracted were two leaden spheres *xx*, each about two inches in diameter, hung from the ends of the rod. The whole was enclosed in a narrow wooden case with glass windows, and the case was supported by levelling screws *SS* on firm uprights let into the ground. By means of an endless screw worked by the rod *K* the torsion head *FF* could be turned from outside so as to make the rod lie centrally in the case.

The attracting masses were leaden spheres *WW*, about 8 inches in diameter, supported by copper rods from a wooden cross rod *rr*. This rod could be turned, also from outside, by means of the rope and pulley *mMM*, so as to bring the weights up to within 1/5 inch from the case opposite the suspended balls, where they were stopped by blocks of wood attached to the walls of the room, or swing them round to the corresponding position on the opposite side. It is obvious that in the two positions the attractions of the large weights on the suspended balls would tend to twist the pendulum from its central position in opposite directions.

Cavendish calculated that the attractions to be observed could not be greater than the 1/50,000,000th part of the weight of the balls, so that a very minute disturbing force would destroy the success of the experiment; and rightly judging that the most difficult disturbance to guard against would be that due to wandering currents of air inside the case caused by unequal heating, he "resolved to place the apparatus in a room which should remain constantly shut, and to observe the motion of the arm from without by means of a telescope." Hence the arrangements for moving the weights and the torsion head from outside.

To observe the deflections a small ivory scale, divided to twentieths of an inch, was set up near each end of the rod; and the rod carried other ivory scales which served as verniers capable of reading the fixed scales to one-fifth of a division : so that he could observe to one-hundredth of an inch and estimate even more closely. The lamps for illuminating the scales and the telescopes for reading them were both placed outside. No other light was admitted to the room.

287. (1) *Steady deflection.*

Let M, m be the masses of the large and small spheres; d the Theory of the distance between their centres when the rod is Experiment. central; $2a$ the length of the deal rod. Let τ be the moment of the couple required to deflect the rod through one radian; θ the circular measure of the angle through which it is deflected when the weights are moved up to one side.

Then
$$\theta \cdot \tau = 2G \cdot \frac{Mm}{d^2} \cdot a \quad \dots\dots\dots\dots\dots\dots(1).$$

(2) *Time of vibration.*

Let I be the moment of inertia of the rod and suspended balls about the axis; and let T be the time of vibration.

Then
$$T = 2\pi \sqrt{\frac{I}{\tau}} \dots\dots\dots\dots\dots\dots\dots(2).$$

Eliminating τ between (1) and (2) we have
$$\frac{4\pi^2 I}{T^2} \cdot \theta = 2G \cdot \frac{Mma}{d^2} \cdot$$

Therefore
$$G = \frac{2\pi^2 I}{T^2} \cdot \frac{d^2}{a} \cdot \frac{\theta}{Mm} \cdot$$

The quantities M, m, d, a are known, and I can be calculated. It only remains to observe θ and T.

288. The following Table contains a typical set of observations taken from Cavendish's paper.

289. The words " negative " and " positive " distinguish those Explanation of positions of the attracting weights in which they the Table. tended to make the rod rest at the lower and higher numbers on the scale respectively.

EXPERIMENT XIV.　May 26.

Weights in negative position.

Extreme points	Divisions	Time			Point of rest	Time of mid. of vibration		
		h.	′	″		h.	′	″
	16·1	9	18	0				
	16·1		24	0				
	16·1		46	0				
	16·1		49	0	16·1			

Weights moved to positive position.

Extreme points	Divisions	Time			Point of rest	Time of mid. of vibration		
27·7	23	10	0	46⎞	—	10	1	1
	22		1	16⎠				
17·3	—	—		—	22·37			
	22		7	58⎞	—		8	5
	23		8	27⎠				
27·2	—	—		—	22·5			
	23		15	2⎞	—		15	9
	22			32⎠				
18·3	—	—		—	22·65			
26·8	—	—		—	22·75			
19·1	—	—		—	22·85			
26·4	—	—		—	22·97			
	23		43	40⎞	—		43	32
	22		44	22⎠				
20	—	—		—	23·15			
	22		49	53⎞	—		50	41
26·2	23		50	37⎠				

Weights moved to negative position.

Extreme points	Divisions	Time			Point of rest	Time of mid. of vibration		
12·4	16	11	7	53⎞	—	11	8	25
	17		8	27⎠				
21·5	—	—		—	17·02			
	17		15	30⎞	—		15	27
	16		16	3⎠				
12·7	—	—		—	16·9			
20·7	—	—		—	16·85			
13·3	—	—		—	16·82			
20	—	—		—	16·72			
13·6	—	—		—	16·67			
	16	11	50	33⎞	—	11	50	58
	17		51	19⎠				
19·5	—	—		—	16·65			
	17		57	53⎞	—		58	6
	16		58	44⎠				
14								

Motion of arm by moving weights from − to + = 6·27.

　　,,　　　　,,　　　　,,　　　,,　　+ to − = 6·13.

　　　　　　Time of vibration at +　　= 7′ 6″.

　　　　　　　,,　　　,,　　−　　= 7′ 6″.

The first column gives the successive turning points as the rod vibrated over the divisions of the scale. From every three successive points the division at which the rod would ultimately come to rest was determined as in the method of weighing by oscillations (§ 29). The results are given in column 4.

Thus $\qquad \frac{1}{2} \left(\frac{27 \cdot 7 + 27 \cdot 2}{2} + 17 \cdot 3 \right) = 22 \cdot 37.$

Columns 2 and 3 give the times of passing the two divisions of the scale (23 and 22) between which the point of rest lay; and from these is calculated the moment at which the rod must have passed the point of rest, $22 \cdot 37$. This is recorded in column 5, from which T, the time of vibration, is found to be $7' \, 6''$.

The ivory scale was $38 \cdot 3$ inches from the centre of motion, and was divided to twentieths of an inch. The circular measure of the angle turned through by the rod when the weights were shifted from the negative to the positive position was therefore

$$\frac{\frac{1}{2} (6 \cdot 27 + 6 \cdot 13) \times \frac{1}{20}}{38 \cdot 3} = \frac{31}{766}.$$

The angle θ was half this.

290. It appears from column 4 that after the weights have been moved to the positive position the point of rest steadily shifts through about a division in an hour towards the upper end of the scale; and the reverse occurs after they have been moved back again. This " creeping " occurred in all the experiments, and Cavendish felt that, as it might indicate a possible source of error, it had to be explained. He tracked it to its source with great ingenuity.

His first idea was that the wire had possibly been twisted slightly beyond its elastic limit, and might gradually take a set, from which it partially recovered on reversing the position of the weights. Accordingly the large weights were moved to the midway position, and the torsion head turned enough to press the suspended balls against the sides of the case, and twist the wire 15 divisions more. But though they were left thus for two or three hours, the rod returned to its natural position, when the torsion head was turned back, and shewed no tendency to creep.

Next, he suspected that the leaden weights and balls might contain traces of iron or other magnetic impurity. In that case, as the rod happened to be set up east and west, they would gradually become feebly magnetized by the earth's field in the direction of the line joining their centres, and thus attract each other magnetically. To test this he made an arrangement by which he could rotate the large weights on the copper rods through half a turn, without entering the room. The weights were moved up to the case overnight, and the pendulum allowed to come to rest. In the morning they were turned halfway round on the copper rods, so as to reverse their magnetic poles. But he could not detect any effect upon the pendulum. The large weights were then replaced by ten-inch magnets which could be turned end for end. This also failed to affect the pendulum. So that there was no trace of magnetic effect either in the large weights or in the suspended balls.

Finally, he concluded that the effect must be due to a difference of temperature between the weights and the air in the case. If the weights were warmer, they would, for some time after being moved up, go on warming the air in the case, thus causing an upward current between the near side and the suspended balls, tending to draw the balls towards the walls of the case. This was tested by setting lamps beneath the weights in the midway position, and leaving them overnight to burn out. On moving the weights up to the case in the morning a much larger creeping effect was obtained. If, on the contrary, the weights were cooled by leaving pieces of ice to melt on them, a large effect was obtained, but *in the opposite direction*. Holes were then drilled in the weights and small thermometers inserted, other thermometers being hung against the case, in such a way that they could all be read by the observing telescopes. It then appeared that there was always a difference of temperature amounting to one or two degrees, between the weights and the air, the weights being the warmer. The effect was thus accounted for satisfactorily.

291. In computing the final results small corrections had to be applied as follows: (1) for the attraction of the weights on the rod; (2) for the attraction of the weights on the farther ball;

(3) for the attraction of the copper rods on the balls and arm; (4) for the attraction of the case on the balls and arm; (5) for the alteration of the attraction of the weights on the balls, according to the position of the arm, and for the effect which that has on the time of vibration. "None of these corrections, indeed, except the last, are of much signification, but they ought not entirely to be neglected." In fact they were most carefully worked out.

From twenty-nine sets of observations values of the density of the earth were obtained ranging from 4·88 to 5·79, with a mean value of 5·448.

292. Since the time of Cavendish the density of the earth has been determined by many observers, some of whom employed modifications of his method, while others used some form of balance, *i.e.* a pendulum oscillating in a vertical plane, the attracting masses being placed beneath it. The results obtained may be summarized as follows:

Date	Observer	Value of Δ
1798	Cavendish	5·448
1837	Reich	5·49
1841–2	Baily	5·674
1849	Reich	5·58
1870	Cornu and Baille	5·5
1878	Poynting	5·4934
1878–81	Von Jolly	5·69
1886	Wilsing	5·579
1895	Boys	5·5270
1896	Braun	5·52725
1898	Richarz and Krigar-Menzel	5·505

293. The student should take an opportunity of referring to Cavendish's original paper, for it would not be easy to find a more superb example of the triumph of patience and accuracy, combined with delicate manipulation, over experimental difficulties that might seem well-nigh insuperable. There is even a lesson to be learned from the curious irony of fate by which, after all his care, Cavendish gave the mean value of his results as 5·48, instead of 5·448, through a mere arithmetical slip in finding the average! Very often in the course of physical measurements the attention is so absorbed in overcoming experimental difficulties, or recording fractions of a millimetre or a second, that far more important considerations are overlooked, or a mistake is made in the number of centimetres or whole degrees. It is not safe in laboratory work to hold by the maxim about taking care of the pence. The pounds must have at least an equal share of attention.

By way of pendant to this proof that men of genius are not infallible, we may cite Newton's celebrated estimate of the density of the earth to shew that their intuitions are sometimes lucky. With little to go upon except the argument that if the earth were lighter than water, it would emerge from the ocean like a cork on one side or the other, Newton proceeds (*Principia*, Book III. Prop. 10), "Since, therefore, the common matter of our earth on the surface thereof, is about twice as heavy as water, and a little lower, in mines, is found about three or four, or even five times more heavy; it is probable that the quantity of the whole matter of the earth may be five or six times greater than if it consisted all of water, especially as I have before shewed that the earth is about four times more dense than Jupiter."

The true value is almost exactly between the limits thus assigned! And this is only one of many guesses thrown out by that commanding genius in all branches of Physics that have since been shewn to be near the truth by the course of modern science.

EXAMPLES.

1. With the following values :

Time of revolution of earth about sun $=365$ d. 6 h. 9 m.
,,   moon ,, earth $=$ 27 d. 7 h. 43 m.
Mean radius of earth's orbit $= 1\cdot487 \times 10^{13}$ cms.
,,   moon's ,, $= 3\cdot84 \times 10^{10}$ cms.

shew that the mass of the sun is about 324,800 times that of the earth.

2. Taking Cavendish's value for the density of the earth, $\Delta = 5\cdot448$, and assuming the earth's mean radius to be $6\cdot37 \times 10^8$ cms., shew that $G = 6\cdot748 \times 10^{-8}$ dynes.

3. Boys found for G the value $6\cdot6576 \times 10^{-8}$ dynes. Shew that this makes $\Delta = 5\cdot52$.......

4. Shew that the mass of the earth is about 6×10^{21} English tons.

ANSWERS TO THE EXAMPLES.

(The numbers in the Examples have not, for the most part, been chosen to give exact answers, but are either taken from actual observations, or such as would occur in practical work. The answers should be worked out *numerically* to an accuracy of about one per cent., by aid of four-place mathematical tables. The student will find the practice in calculation quite worth the trouble. Such forms as $\pi\sqrt{2}$, $\epsilon^{2\pi}$, &c. should not be left in an answer, except for a special reason, but should be evaluated by the Tables.

Unless the contrary is stated, g may be taken $=32$, and tons are English tons of 2240 lbs.)

Chapter II., p. 21.

1. 100 lbs.   **2.** 7·854 in.   **3.** 270 lbs.   **4.** 2880 lbs.
5. 240 lbs.   **6.** 60 lbs.   **7.** (1) 14 lbs., (2) $21\frac{1}{2}$ lbs.

Chapter III., p. 30.

3. 3 ft. 4 in. from weight 2.   **4.** $a\sqrt{3}$.   **5.** $\frac{3}{26}$ in.
6. $2\frac{5}{8}$ in. from centre.   **7.** 2685·2 inches from centre of earth.
9. $3\frac{3}{7}$ in. from base.   **10.** 16 lbs. ; 8 feet.

Chapter IV., p. 39.

1. The arms are unequal. 20·599 gms.   **2.** 38·5 mgms.
4. 11 lbs. $10\frac{2}{11}$ oz.   **6.** $\dfrac{Q^3}{P^2}$.   **7.** ·058 in.

Chapter V., p. 47.

1. 8800 lbs.   **2.** 390 lbs.   **3.** $5 : \sqrt{41}$.

Chapter VII., p. 63.

1. 3·73 lbs.   **2.** $\dfrac{l}{2\pi}\cot a\left(1+\dfrac{W}{2\pi T_0}\cot a\right)$.

CHAPTER IX., p. 83.

1. 196,384 miles per sec. **2.** 1 mile 1226 yds. 2 ft. **3.** 88.
4. (1) $17\frac{1}{22}$; $20\frac{5}{11}$. (2) $18\frac{3}{4}$. (3) 48. (4) 1 : 507. **5.** 11·18 sec.
6. 324 ft. **7.** 256 ft.

CHAPTER X., p. 87.

1. 1 : 289. **2.** 25 : 9. **3.** 12·9 ft. sec. units.

CHAPTER XII., p. 114.

1. 22; $29\frac{1}{3}$; $52\frac{4}{5}$; $5\frac{5}{11}$; $7\frac{1}{2}$; $27\frac{3}{11}$. **2.** 210670.
3. 18·391 miles per sec. **4.** 1037·5 miles an hour.
5. 27,777·$\dot{7}$: 911·4. **6.** 774·4 yds. ; 25·2 mls. an hr; $62\frac{6}{9}$ sec.
7. 1717·3 yds. **8.** 1098 ft. per sec. **9.** $\frac{11}{15}$ ft. sec. units.
10. 7·5. **11.** 112; 208; 400. **12.** 768.
13. After $3\frac{3}{4}$ sec.; 24 upwards; 8 down; 104 down. **14.** 1·79.
15. 1·955; 220 ft. **16.** (a) 320,000; (b) 270,000.
17. $1\frac{1}{2}$ sec.; 6; 510; at the end of the ten seconds. **18.** $u + \frac{a}{2}(2n - 1)$.
19. 355 days 5 hrs; 2854170 million miles. **20.** 100 sec.; 40,000 ft.
21. 8163 metres. **22.** -2; -2; 5 sec.; $12\frac{1}{2}$ feet.

CHAPTER XIII., p. 134.

3. (1) (a) 6; (b) 4; (2) (a) 125; (b) 222·46. **4.** 24; 9; 1344; $\frac{3}{4}$; $\frac{9}{32}$; 42.
5. 16326·5. **6.** 96 lbs.; $2\frac{2}{3}$ lbs.; 1 lb.; 50 gms.
7. 6673·8; 33369 cms. **8.** ·24; (a) 7·2; (b) 35·6; 10800 ft.; 4 m. 56·6 sec.
9. 3·094. **10.** 75·3; 400 ft.; $27\frac{3}{11}$ sec. **11.** $4\frac{7}{12}$ tons; $1\frac{7}{48}$ tons.
12. 625 lbs. **13.** 44·64 ft. per sec. **14.** 1350000; 1318·3 lbs.; 3·6 in.
15. ·12; ·1. **16.** 9·76 lbs. **17.** 17 times. **18.** 358 lbs.
19. About 7° 17'. **20.** 43·86 tons.

CHAPTER XIV., p. 141.

1. 57·2; 457·8. **2.** 980·4. **3.** $x = 15$ gms.; $g = 978·9$.
4. $5\frac{1}{3}$; $10\frac{2}{3}$ ft. **5.** 224·91 gms.; 223 gms.; $1\frac{2}{3}$ lbs.

CHAPTER XV., p. 153.

1. (a) 82665779; (b) 81312000. **2.** 89 : 1. **4.** $\frac{14}{165}$ H.P.
5. $4\frac{1}{6}$. **6.** $29\frac{19}{64}$. **7.** ·10625. **8.** 134·4.
9. 1469 yds.; $34\frac{2}{9}$ miles. **10.** 24·42 lbs. **11.** $18\frac{3}{4}$ lbs.
12. 57600 lbs. **13.** ·00014178.... **14.** 107·4. **15.** 450,000 lbs.
16. About 4,940,000. **17.** 168·19 lbs.

Chapter XVI., p. 158.

1. 61° 24′.　　　　**2.** 8·8 in.　　　　**3.** 182180 miles per sec.

Chapter XVII., p. 169.

3. 60° from smaller force.　　　　**8.** (1) 25 lbs.; 16° 16′ from 7 lb. force.
(2) 13 lbs.; 27° 48′ from 8 lbs.　　(3) 12·76 lbs.; 48° 15′ from 14 lbs.
(4) 12·61 lbs.; 22° 1′ from 8 lbs.　　**9.** 18·00 lbs.; 70° 40′ from 7 lbs.
10. 9·66 lbs.; 96° 8′ 36″ from force 1 on opposite side to force 2.
11. 45·96 lbs.　　　　**12.** 103·92 lbs.　　　　**13.** 500 lbs.; 300 lbs.
14. $20\frac{5}{6}$ lbs.; $54\frac{1}{6}$ lbs.　　**15.** 9 lbs.; 12 lbs.　　**19.** 36·6 lbs.; 25·8 lbs.
20. 4·62 lbs.; 8 lbs.　　**24.** $\tan\theta = \dfrac{P}{Q}$; $T = \dfrac{PQ}{\sqrt{P^2 + Q^2}}$.
25. $37\frac{1}{2}$ lbs.　　**26.** 62° 42′; 11 lbs.

Chapter XVIII., p. 189.

3. $\sqrt{P^2 - W^2}$.　　　　**4.** 20·8 lbs.; 70° 54′ to horizon; 158 lbs.
6. 70° 53′ with horizon; 26·45 lbs.
8. $\tan\theta = \dfrac{a\cot a - b\cot\beta}{a+b}$; $W\dfrac{\sin\beta}{\sin(a+\beta)}$; $W\dfrac{\sin a}{\sin(a+\beta)}$.
9. Let W be the weight of the rod; θ its inclination to the vertical. Then
$$T = \frac{W}{2\cos a}; \qquad \cot\theta = \frac{a-b}{a+b}\cot a,$$
where $\qquad\qquad \cos a = \dfrac{a+b}{2}\sqrt{\dfrac{l^2-(a+b)^2}{ab}}$.
10. $W\tan a$; $W\sec a$.　　**12.** 30° to vertical; 11·55 lbs.; 5·77 lbs.
15. 28·3 lbs.　　**16.** 91 lbs.　　**17.** $102\frac{2}{3}$ lbs.; 0.　　**18.** 20 lbs.
20. $4\sqrt{2}$.　　**21.** 6 lbs. along DA produced; 18·186 AB.

Chapter XIX., p. 203.

1. 16·8 lbs.; 18·277 lbs.　　**2.** ·25.　　**3.** 6 tons.
4. ·115; 30°.　　**5.** 791 : 1000.　　**6.** ·577.　　**8.** About $12\frac{1}{2}$ tons.

Chapter XX., p. 211.

1. 200 ft.; 10 sec.　　**2.** $12\frac{1}{2}$ sec.; 34 miles an hour.
3. About 16 sec.; 26 miles an hour; 450 feet.　　**4.** 8 ft. per sec.
7. 2·23 sec.　　**8.** 30°.　　**9.** 6 oz.　　**10.** 37 ft. 6 in.

CHAPTER XXI., p. 222.

1. 2·5 sec. ; 173·2 ft. ; 25 ft.   **2.** 29·7 miles.   **3.** 2000 ft. per sec.
4. 38° 19′ or 85° 26′.   **5.** 20·057 miles.   **6.** 109·75 ft. per sec.
7. 176·7.   **8.** 86·4 ft. per sec. ; 170·64 ft. per sec.
10. 16·4 ft. per sec. ; 31 ft. 0¾ in.   **11.** 72·2 sec. ; 15·8 miles.
13. 37·02 ft. per sec.

CHAPTER XXII., p. 235.

1. (1) 1·57 sec. ; (2) 1·256 sec. ; (3) 1·57 sec.   **2.** (1) 4 ft. per sec. ;
(2) 50 cm. per sec. ; (3) 8 ft. per sec.   **4.** 2·09 sec. ; 117 ft./sec².
5. 27.   **6.** ·4014 sec. ; 156·53 cm./sec.   **7.** 1,293,600 dynes.
8. 2·5 sec. ; 3·02 sec.

CHAPTER XXIII., p. 241.

1. 9·78496 in.   **2.** 34″.   **3.** 1/160 in.   **4.** 15·2 cm.
5. 32·197.   **6.** 5·13.   **7.** 10·8.   **8.** 3666·6 ft.   **9.** 1098·8 ft.

CHAPTER XXIV., p. 255.

5. 4332·5 days ; about 480 million miles.

CHAPTER XXV., p. 267.

1. (1) 4$\frac{2}{7}$ ft. per sec. ; 5$\frac{5}{7}$ units.   (2) $v_1=4\frac{3}{35}$; $v_2=5\frac{2}{7}$; 9$\frac{1}{7}$ units.
2. $v_1=-4$; $v_2=5\frac{1}{3}$; 93$\frac{1}{3}$ foot-poundals.   **5.** 1$\frac{1}{44}$ sec. ; ·88 inch per second.
6. 331 lbs.   **7.** 51400 lbs.   **8.** 3 blows.
10. 60° 57′ 18″ ; 29° 2′ 42″.   The directions are at right angles.
11. 44·6 ft. per sec.   **12.** The axis is diminished by six-millionths,
i.e. about 550 miles.   The year is shortened by 4 min. 24 sec.

CHAPTER XXVI., p. 275.

1. 6·26 inches.   **2.** 58·71 inches.

CHAPTER XXVIII., p. 287.

1. 274.   **2.** 17·4 tons.   **3.** 348·52 ; 31 secs.
4. (1) 1·69 ft./sec² ; (2) 1·788 ft./sec².   **5.** 64·15.

6. (Complete oscillation) $2\pi \sqrt{\dfrac{a\sqrt{2}}{3g}}$; 10·37 inches.   **7.** 116·25.

CHAPTER XXIX., p. 302.

1. 243·515.   **2.** 789·4 ft. per sec.

CHAPTER XXX., p. 312.

1. 32·1906.

INDEX.

The numbers refer to the pages.

For EU product safety concerns, contact us at Calle de José Abascal, 56–1°,
28003 Madrid, Spain or eugpsr@cambridge.org.

www.ingramcontent.com/pod-product-compliance
Ingram Content Group UK Ltd.
Pitfield, Milton Keynes, MK11 3LW, UK
UKHW010852090126
466816UK00011B/176